NUCLEAR MEDICINE PHYSICS

Edited by **Aamir Shahzad** and **Sajid Bashir**

Nuclear Medicine Physics
http://dx.doi.org/10.5772/intechopen.75255
Edited by Aamir Shahzad and Sajid Bashir

Contributors

Khaled Soliman, Ahmed Alenezi, Abdullah Alrushoud, Salman Altimyat, Hasna Albander, Turki Alruwaili, Ya-Yao Huang, Sajad Bashir, Aamir Shahzad, Geofery Luntsi, Victory S. Daniel, Chigozie I. Nwobi, Paul T. Bura, Meenu Gupta, Anupam Dhasmana

First published in London, United Kingdom, 2019 by IntechOpen
IntechOpen is the global imprint of INTECHOPEN LIMITED, registered in England and Wales, registration number: 11086078, The Shard, 25th floor, 32 London Bridge Street
London, SE19SG – United Kingdom
Printed in Croatia

British Library Cataloguing-in-Publication Data
A catalogue record for this book is available from the British Library

Additional hard and PDF copies can be obtained from orders@intechopen.com

Nuclear Medicine Physics, Edited by Aamir Shahzad and Sajid Bashir
p. cm.
Print ISBN 978-1-83880-283-7
Online ISBN 978-1-83880-284-4
eBook (PDF) ISBN 978-1-83880-500-5

For EU product safety concerns:
IN TECH d.o.o., Prolaz Marije Krucifikse Kozulić 3, 51000 Rijeka, Croatia,
info@intechopen.com or visit our website at intechopen.com.

Meet the editors

Aamir Shahzad has over 14 years of experience of university research and teaching at home and abroad, and due to his continuous efforts the thermophysical properties of materials are now available at graduate and postgraduate levels. Dr. Shahzad received his Post Doctorate and Doctoral degrees from Xi'an Jiaotong University, P.R. China, in 2015 and 2012, respectively. He has proposed novel methods to explore the outcomes of complex materials, which show his aptitude to comprehend computational physics and modeling and molecular simulations together with experimental understanding. Moreover, Dr. Shahzad is currently working in the fields of computational physics and complex fluids/plasmas, in addition to computational nuclear medicine and bio- and energy materials in collaboration with Dr. Sajid Bashir (Principal Scientist, PINUM Cancer Hospital, Pakistan).

Dr. Sajid Bashir received his Doctoral and MSc degrees in Physics from the State University of New York at Albany, USA, in 2015. He received his MSc (Medical Physics) through a fellowship program from the Pakistan Institute of Engineering and Applied Sciences, Islamabad, Pakistan, which is a top-ranking institute in Pakistan. He was successful in revealing the phase contrast effect by employing a specialized glass tube in a technique called polycapillary optics. He has published more than 10 international research papers in well-reputed journals and presented his research findings in international and national conferences. Currently, Dr. Sajid is working as a senior medical physicist at PINUM Cancer Hospital and as an adjunct faculty at the Department of Physics, GC University Faisalabad (GCUF), since 2003. Dr. Sajid is a member of a number of international and national scientific and research organizations, including the American Association of Physicists in Medicine, the Pakistan Society of Nuclear Medicine, and the Radiological Society of Pakistan. He is also a recipient of a National Institutes of Health award during his doctorate studies in the USA.

Contents

Preface

This book covers all aspects of nuclear medicine production of radionuclides and their uses for diagnosis, treatment, and instrumentation. Nuclear medicine is a powerful translational tool in the basic sciences, such as biology, in drug discovery, and in preclinical medicine. Improvements in nuclear medicine are motivated by progress in this multidisciplinary science, which includes physics, chemistry, computing, mathematics, pharmacology, and biology. The book has been written for undergraduate and postgraduate students of medical physics who want to make the foundation of knowledge in this field stronger. It also serves as a resource for interested readers from other disciplines, for example, clinicians, radiochemists, and medical technologists who would like to familiarize themselves with the basic concepts and practice of nuclear medicine physics. This book addresses an urgent need for a comprehensive, contemporary text on the physics of nuclear medicine and aims to fill the knowledge gap between the available research concerning nuclear medicine and basic medical physics. The physics of nuclear medicine is explained in detail and, wherever possible, the physical interpretations are explained. The book's clarity, in terms of research, and completeness make it suitable for self-learning and for self-paced objectives. Here is a quick run-through of the basics: In the introductory chapter, we explain the discussion of nuclear medicine that involves the administration of trace quantities of radionuclides used to provide diagnostic information in a diverse range of diseases. The second chapter incorporates radiation therapy in metastatic neuroblastoma. The role of radiotherapy as a palliative modality in patients with advanced neuroblastoma provides better symptomatic relief. The third chapter covers low-dose radiation-induced effects on white blood cell counts in guinea pigs. The fourth chapter addresses the positron emission tomography (PET) radiopharmaceuticals listed in the US Pharmacopeia (USP) or European Pharmacopeia (EP). PET radiopharmaceuticals listed in monographs of the latest USP and/or EP are included in this chapter. The fifth chapter tackles radiation protection in the routine practice of both diagnostic and therapeutic applications in nuclear medicine, including PET, diagnostic facility design, and safety aspects of common radionuclides used in clinics. The last chapter presents the development of diagnostic reference levels of standard doses in nuclear medicine.

Dr. Aamir Shahzad
Assistant Professor
Molecular Modeling and Simulation Laboratory
Department of Physics
Government College University Faisalabad
Pakistan

Chapter 1

Introductory Chapter: Role of Nuclear Medicine in Medical Science

Aamir Shahzad

Additional information is available at the end of the chapter

http://dx.doi.org/10.5772/intechopen.85606

1. Introduction

The science of nuclear medicine (NM) involves the administration of trace quantities of radionuclide's that are used to provide diagnostic information in a diverse range of diseases. In its most basic form, a NM study comprises of injecting a radiopharmaceutical, a combination of specific pharmaceutical tagged with a gamma-ray-emitting radioactive tracer into the body. There are a number of pharmaceutical available which are used for specific organ imaging. It function is to carry gamma emitting radioisotope into a specific organ. When the radionuclide decays, gamma rays photons are emitted. The energy of these gamma photons is such that a large number of photons are exited from the body without being scattered or attenuated. These photons are later detected by a position-sensitive instruments called gamma camera or scintillation camera and form an image of the distribution of the radionuclide, and hence the compound to which it was attached. There are two classes of nuclear medicine imaging: single photon emission tomography which is essentially a single photon imaging and positron imaging. Single photon imaging usually comprises of either taking a planar image or a series of planar images around the body. A planar image is picture of radionuclide distribution in the patient from one angle. This results in an image having insufficient depth information, but which can still be diagnostically useful. In order to get depth information, data from various views are collected around the patient. This allows cross-sectional images of the distribution of the radionuclide which was later reconstructed employing specialized soft ware's (these software's use highly sophisticated algorithms), thus providing the depth information missing from planar imaging. Positron imaging uses radionuclide that decay by positron emission. The emitted positron usually has a very short lifetime and produces two high-energy photons after interacting with its counterpart electron. The two simultaneously emitted gamma photons having energies of 511 KeV subsequently are detected by an imaging

Material	Density	Atomic number (Z)	μ@511 keV (1/cm) (mean free path-mm)	Photo fraction (%)	Light output (photon/MeV)	Decaytime (ns)	λ (nm)	Energy resolution (%FWHM)	Hygroscopic	Index of ref
BGO	7.1	75	0.95 (10.5)	40	9000	300	480	12	N	2.15
GSO	6.7	59	0.70 (14.1)	25	8000	60	440	9	N	1.85
LSO	7.4	66	0.88 (11.4)	32	30,000	40	420	10	N	1.82
NaI (TT)	3.7	51	0.34 (29.1)	17	41,000	230	410	8	Y	1.85

Table 1. Physical properties of PET scintillators.

camera. Once again, topographic images are formed by collecting information from different angles around the patient, resulting in PET images; however detectors remain stationary and do not move around the patient as it happens in SPECT study. A list of physical properties of different types of scintillator used in nuclear medicine is described in **Table 1**. It is clear from the table the GSO and LSO are quite fast materials with the decay time of 60 ns and 40 ns and are most suitable for PET time of flight measurements. However, NaI (Tl) detectors are more sensitive and give strong output per unit absorption of energy.

Although the amount of radioactive tracers injected into the patient is very small, however, smaller quantities carry risk and therefore assessment of absorbed dose to the organs and whole body become essential. Considerable work has been done internationally so far on the assessment of dose to remnant thyroid tissue and whole body. A high level of radioactivity is usually prescribed in routine to ablate thyroid tissue, therefore, its accurate quantification as well as safety of radiation technologists is a must. Different methods of measuring activity while in shielding were proposed to reduce the extra radiation burden to allied radiation staff. Before giving therapeutic radioiodine, uptake in the thyroid tissue is determined using uptake system that provides an estimation of the remnant thyroid mass. The uptake value is a value that can be reproduced with great accuracy and same will also be made sure that the uptake value is reproducible. Currently the clinical practice of administering radioactivity to treat Differentiated Thyroid Cancer (DTC) varies widely from hospital to hospital and ranges from 1110 to 3700 MBq or even more. This increase in activity although does not confer any therapeutic benefits to the patients. The corresponding absorbed dose to thyroid mass also varies widely from (13–1161) Gy depending on the mass of the remnant thyroid tissue, dose rate and the absorbed cumulated activity. The whole body receives, in this case, an absorbed dose of 0.12–0.23 Gy. Since the radioiodine is a non-specific agent, it also deposits in other parts of the body giving unnecessary radiation dose for example breast, liver, etc. This is particular of important for lactating women. The empirically determined activity without any apparent correlation between absorbed dose and activity depends solely on the experience of the individual groups and can varies by an order of magnitude from the standard practice. High success of non-scientific approach was reported, however, 15% of patients with high-risk DTC have significantly reduced life expectancy even after getting treatment using conventional approach of fixed amount of administered activity. The fixed-activity approach without assessing pretherapeutic lesion absorbed dose and toxicity assessment generally results in administration of low amount of therapeutic radiodine as compared to with absorbed dose assessment. In this era of personalized and precision medicine, individualized approach to treatment will bring more patient benefits and improve life expectancy. The quantity of activity should be given to patient that is right and as high as safely achievable (AHASA) [1–5].

2. Conclusions

Diagnostic reference levels (DRLs) and achievable doses (ADs), a form of investigation levels, represent an important tool in medical imaging as optimizing the radiation dose delivered to patients. It is essential to ensure that the appropriate clinical information is available in the image throughout the optimization process. In order to implement optimization process, both patient dose and clinical utility must be taken into account depending on image quality.

Acknowledgements

The authors thank Dr. Sajid Bashir (Principal Scientist, Punjab Institute of Nuclear Medicine (PINUM) Faisalabad, Cancer Hospital, Pakistan) for providing his useful discussions and suggestions.

Abbreviations

NM	nuclear medicine
SPECT	single photon emission computed tomography
GSO	gadolinum orthosilicate
LSO	lutetium orthosilicate
PET	positron emission tomography
NaI(Tl)	sodium iodide thallium activated
MBq	mega becqural
DTC	differentiated thyroid cancer
Gy	gray
AHASA	as high as safely achievable

Author details

Aamir Shahzad

Address all correspondence to: aamirshahzad_8@hotmail.com

Molecular Modeling and Simulation Laboratory, Department of Physics, Government College University Faisalabad (GCUF), Faisalabad, Pakistan

References

[1] I.C.R.P. 1990 recommendations of the international commission on radiological protection. Annals of the ICRP. 1991;**21**(1-3):123-127

[2] I.C.R.P. International Commission on Radiological Protection. Radiological protection and safety in medicine. Annals of the ICRP. 1996;**26**(2):1-31

[3] N.O.H.S.C. National Occupational Health and Safety Commission. Recommendations for Limiting Exposure to Ionising Radiation. Oxford, England: Pergamon; 1995

[4] Toosi BMT, Azadinezhad M. Local diagnostic reference levels for some common diagnostic X-rays examinations in Tehran County of Iran. Radiation Protection Dosimetry. 2007;**124**(2):137-144

[5] Huda W, Nickoloff EL, Boone JM. Overview of patient dosimetry in diagnostic radiology in the USA for the past 50 years. Medical Physics. 2008;**35**(12):5713-5728

Chapter 2

Radiation Therapy in Metastatic Neuroblastoma

Meenu Gupta and Anupam Dhasmana

Additional information is available at the end of the chapter

http://dx.doi.org/10.5772/intechopen.80892

Abstract

Neuroblastoma is the commonest extracranial solid tumor in children, and metastasis at presentation is seen in more than 50% of cases. The role of radiotherapy as a palliative modality in patients with advanced neuroblastoma provides better symptomatic relief. Palliative radiotherapy dose schedules can be given either in single hypofractionation from 4 to 8 Gy or fractionated radiotherapy that can range from 21 to 30.6 Gy. Dose-response relationship trend has been reported in the palliative setting of bone metastasis. Because of the proximity of tumor to critical organs, serious adverse effects can be avoided with conformal techniques. Although currently there is limited data available, new treatments with particle therapies are undergoing clinical evaluation and may offer new hope for good quality of life in these patients.

Keywords: neuroblastoma, radiotherapy, palliative, metastasis, dose

1. Introduction

Rudolf Virchow, a German physician, was the first person who described neuroblastoma in 1864. Through his research he called the tumors found in the abdomens of children as gliomas. In 1910, James Homer Wright made great efforts to map the origin and development of tumor cells, and he named the tumors neuroblasts, as "blastoma" refers to a collection of immature, undifferentiated cells [1]. His meticulous study showed that these tumors originated from an immature, primitive form of neural cell. The term *neuro* denotes to "nerves," while *blastoma* denotes to a "cancer that affects immature or developing cells." Neuroblasts which are formed during embryonic stages develop normally once fetus matures after birth. Sometimes due to uncontrolled cell divisions, they become cancerous, causing neuroblastoma (**Figure 1**).

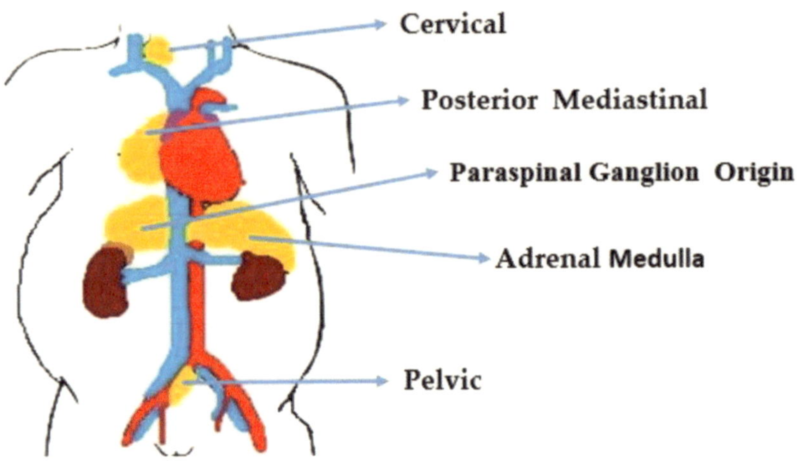

Figure 1. Primordial neural crest cells which develop into the adrenal medulla and sympathetic ganglia during specification and differentiation that lead to formation of neuroblastoma. Adrenal gland is the most common site of origin, but sympathetic nervous system chain from the neck to the pelvis can be other common site of origin.

2. Epidemiology

This set of tumors is considered almost exclusively a disease of pediatric population. Neuroblastoma is the most common solid extracranial tumor, accounting for third most common childhood cancer, after leukemia and brain tumors. Neuroblastoma accounts for approximately 15% of all pediatric cancer fatalities with almost 600 new cases are diagnosed in the United States each year. White race infants displayed more incidence of neuroblastoma than black infants (ratio of 1.7:1 for male and 1.9:1 for females), but little of any racial difference is apparent among older children. Males have higher incidence rate relative to females (male-to-female ratio of 1.3:1) [2]. Neuroblastoma is thought to occur sporadically, with 1–2% of cases are familial [3].

3. Clinical presentation

Patients with neuroblastoma present with a combination of signs and symptoms which are nonspecific and are variable. They depend on site of tumor, size, degree of metastatic spread, and catecholamine secretion.

3.1. Primary disease

Neuroblastomas in the abdomen are the most common form that arises in 65% of cases, approximately 50% arises from the adrenal glands (adrenal neuroblastoma), and one-third are from paravertebral ganglion (extra-adrenal neuroblastoma). Abdominal mass which is asymptomatic may be an incidental finding which is usually detected by parents [4].

Symptoms are abdominal pain or fullness and abdominal mass which is usually firm and fixed. Characteristic nodularity of abdominal neuroblastomas is similar to palpating a *bag of potatoes* which differentiates it from a nephroblastoma. In some cases, enlarged liver, spleen, and intestinal obstruction may be seen. Rarely, scrotal and lower extremity edema results from the compression of venous and lymphatic drainage of the lower extremities due to size of primary or metastatic abdominal tumors. Renin-mediated hypertension is because of compromised renal vasculature. Epinephrine is rarely released from most neuroblastomas due to deficiency of enzyme necessary for synthesis. So hypertension, tachycardia, flushing, and sweating are uncommon symptoms. Hypertension and opsomyoclonus detected in less than 2% are part of paraneoplastic syndromes [5].

3.2. Metastatic disease

More than 50% of patients present with metastatic disease. Neuroblastoma is associated with lymphatic and hematogenous spread. Common hematogenous metastatic sites are the bone marrow (70.5%), bones (skull, long bones, ribs, vertebrae, 55.7%), liver (29.6%), and skin and rarely the brain or lungs. Thirty-five percent of patients had regional lymph node metastases with localized tumors [6].

3.2.1. Signs and symptoms of metastasis

Long bone involvement causes pain and limping with increasing risk of pathological fractures which is known as Hutchinson syndrome. Periorbital ecchymoses, swelling, and proptosis (raccoon eyes) occur due to sphenoid bone and retrobulbar tissue involvement. Painless proptosis, periorbital edema, and ecchymosis of the upper lid are akin with trauma or child abuse. Irritable and fussy infant needs attention. Bone marrow involvement may result in pancytopenia. Huge involvement of the liver in metastatic condition is common in infants with stage Ms/4S and may cause Pepper syndrome (PS), which induces the respiratory distress, quoted by W. Pepper [7]. PS was recognized as a localized primary tumor and metastatic condition restricted to the skin, bone marrow, and liver in infants. PS is usually known for better prognosis, as it is linked with spontaneous regression of disease. Some infants with stage 4S neuroblastoma die of massive hepatomegaly, respiratory failure, and overwhelming sepsis. Non-tender, bluish, and mobile skin and subcutaneous nodules are because of metastasis to these sites. This is known as "blueberry muffin sign." These nodules become

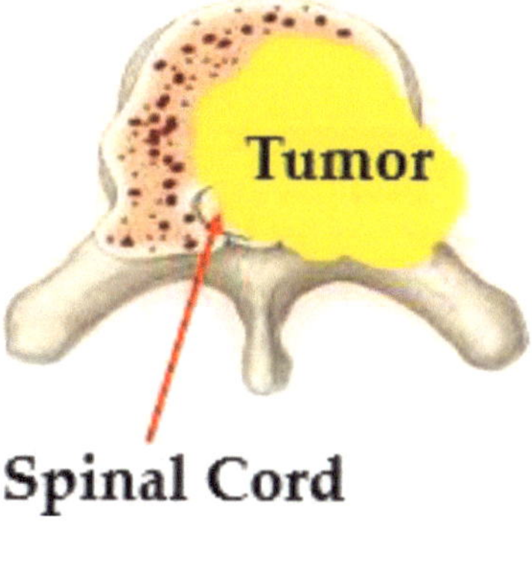

Figure 2. Spinal cord compression by vertebral body metastasis and extension of tumor posteriorly into the epidural space.

prominently red once provoked and thereafter blanch for some minutes, due to release of vasoconstrictive metabolic products. These nodules can be diagnostic of neuroblastoma and should be differentiated from metastatic skin leukemic infiltrates.

Paraspinal tumors in the thoracic, abdominal, and pelvic regions often present with spinal cord compression due to spinal canal invasion through the neural foramina causing symptoms related to compression of nerve roots and spinal cord. There may be subacute or acute paraplegia, bladder or bowel dysfunction, or less commonly radicular pain. Cervical neuroblastoma may present as Horner's syndrome [8] (**Figure 2**).

4. Prognosis according to MYCN amplification status

In infants, MYCN (neuroblastoma-derived, v-myc avian myelocytomatosis viral related oncogene) amplification and unfavorable Shimada histopathology correlated with increased frequencies of bone and intracranial or orbital metastases. In geriatric patients, MYCN amplification is associated with increased risk of intracranial or orbital and lung and liver metastases [6]. MYCN amplification status defines response rate to palliative radiotherapy. Median overall survival time is increased in patients without amplification. A French study showed response rate to palliative RT according to MYCN status was 47.6% in patients with amplified MYCN vs. 75.7% in patient without amplification of MYCN ($p = 0.04$) [6, 9].

5. Investigation

Medical history including family history and physical examination is needed before proceeding for any investigations.

- Laboratory studies: urine examination for catecholamine homovanillic acid (HVA) and vanillylmandelic acid (VMA)
- Imaging: abdominal ultrasound and computed tomography/MRI (calcification on imaging is a favorable sign)
- MIBG (metaiodobenzylguanidine) scan, bone scan, and positron emission tomography (PET) scan
- Biopsies and bilateral bone marrow biopsy

6. Staging

Most commonly used system is the International Neuroblastoma Staging System (INSS) based on clinical, radiographic, and surgical findings (see **Figure 3**).

http://dx.doi.org/10.5772/intechopen.80892

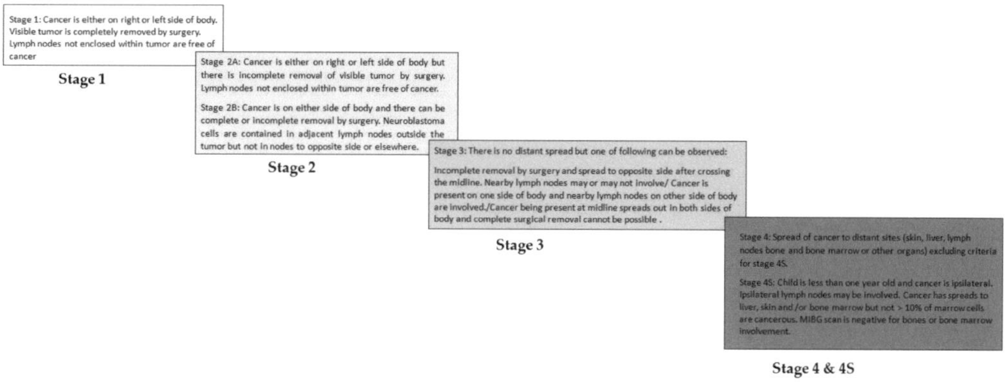

Figure 3. The International Neuroblastoma Staging System (INSS) [10, 11].

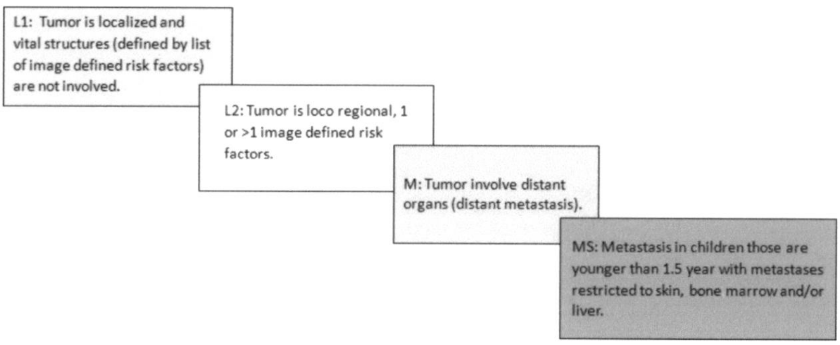

Figure 4. Showing the International Neuroblastoma Staging System (INRGSS) [10, 11].

Later, the International Neuroblastoma Risk Group Staging System (INRGSS) and International Neuroblastoma Risk Group Consensus Pretreatment Classification are released. INRGSS is using pretreatment tumor imaging rather than extent of surgical resection [10]. The INRGSS is explained in **Figure 4**.

7. Management

1. For stage I and II disease, the treatment is complete removal of both the primary tumor and its adjacent involved lymph nodes. Cytoreductive removal of the maximum tumor burden can be done in more advanced stage with expertise of oncosurgeons.

2. Surgical resection is not recommended in stage 4S disease where the neuroblastoma is prone to spontaneous regression.

3. Multidisciplinary approach with surgery, chemotherapy, and irradiation is a key for neuroblastoma treatment depending on patient age, disease stage, response to therapy, and tumor relapse. Local control, metastatic control, and the prevention of relapse are the goal of treatment

7.1. Radiotherapy

In high-risk neuroblastoma, radiation therapy (RT) can be recommended both to the primary site and to sites of metastatic disease as part of consolidative therapy. RT also plays an integral part in the palliation of symptoms from metastatic disease. RT can be used to treat emergencies such as:

- Cord compression
- Tracheal compression
- Expanding retro-orbital tumor
- Imminent bone fracture
- Rapidly enlarging liver

7.1.1. Radiation therapy techniques

7.1.1.1. Three-dimensional conformal radiation therapy

Three-dimensional conformal radiation therapy (3DCRT) is considered the gold standard practice to treat metastases. Conformal dose distribution to the target volume and dose reduction to the surrounding normal tissues is achieved by 3DCRT. In 3DCRT planning, after making the patient comfortable and in reproducible position, adequate immobilization is done by immobilization devices. Three-dimensional anatomic information is obtained on CT simulator or MRI. Three-dimensional conformal plans are generated to deliver high doses within the tumor which is the target and spare the adjacent normal structures at the same time. In the case of vertebral metastasis, radiation portals should include the involved vertebral body (and the soft tissue if involved by the tumor), plus a vertebral body below and above. Patients can be treated in either supine or prone position. If bony metastatic sites are very painful and there is difficulty in prone position, patient can be made comfortable in supine position, and this is also an acceptable reproducible position. A three-dimensional plan composed of one anteroposterior field and two posterior oblique wedged field is illustrated in **Figure 5**. The dose prescribed is 20 Gy at 4 Gy per fraction. Hypofractionated radiotherapy is minimally toxic and time efficient for palliation purpose.

7.1.2. Treatment volumes

Delineation of target volumes is done by CT-based planning. The gross tumor volume of the primary (GTVp) should include post-induction chemotherapy and presurgical disease. For abdominal primaries, the clinical target volume (CTVp) included the para-aortic lymph nodes in addition to the GTVp. Setup uncertainties are reduced by the generation of planning target volume (PTVp) through expansion of the CTVp approximately 0.5–1.0 cm. For metastatic sites irradiation, the GTVm consisted of the residual metastatic tumor (following induction chemotherapy) as defined by MIBG, CT, or MRI. An additional 1.0–1.5 cm CTVm margin for microscopic disease followed by a 0.5–1.0 cm PTVm margin will account for setup uncertainties.

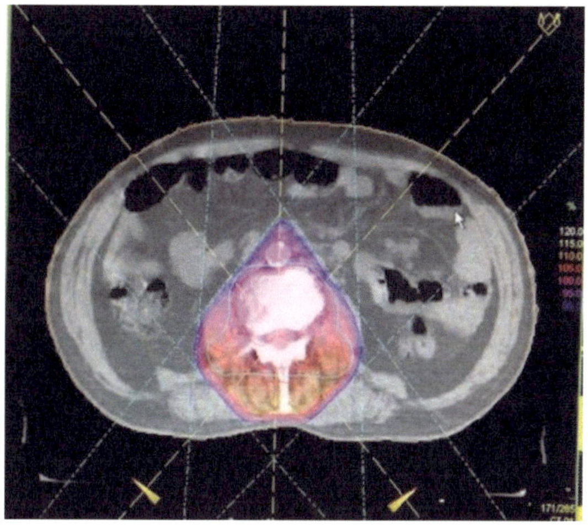

Figure 5. Isodose distribution for three-dimensional conformal radiotherapy (3DCRT) to vertebra.

The 6MV photon energy is delivered by linear accelerator. Wedges and compensators are used to make the dose distribution more uniform. Whole PTV should be included within the 95% isodose surface. Not >10% of the PTV should achieve >110% of the prescription dose (evaluated by DVH) [12].

If dose prescription is 21.6 or 36 Gy, then constraints defined by ANBL0532 AOR are the contralateral kidney in which V12 should be <20% and V8 < 50% and the liver in which V9 should be <50% and <25% to receive 18 Gy. With these conservative doses, Kandula et al. did not observed any hepatic or renal toxicity [12].

When treating a child with intensity-modulated radiotherapy (IMRT) techniques, the theoretical increase in secondary malignancies should be considered. For volumes exposed to low-dose RT, IMRT significantly increased the normal tissue volume receiving 50% or less of the prescribed dose for the volumes exposed to low-dose RT. The decision to irradiate a metastatic site should be made by the treating physician depending upon the evaluation based on post-induction response and posttransplant or preradiotherapy imaging.

7.1.3. Radiotherapy doses

Most of the studies recommend that the median RT dose to the primary site is 21.6 Gy (range 21.0–30.6 Gy). Metastatic sites can be irradiated concurrently with the primary site [13]. The median RT dose recommended to metastatic sites is 21.6 Gy (range 21.0–30.6 Gy) in 12 fractions [12]. Casey et al. reviewed results of RT to metastatic sites as a part of consolidative therapy at a single institution between 2000 and 2015. Among 159 patients, 229 metastases were irradiated. Median dose of 21 Gy (10.5–36 Gy) was given to 90% of irradiated metastasis. Out of 229 irradiated metastasis, 35(15%) had tumor recurrence. These irradiated metastatic sites had 81.3% 5-year local control. There was no difference in local control as far as number of metastatic sites is irradiated. Also the site of metastatic bone versus soft tissue irradiation had no impact on local control, but there was improved overall survival (OS) 59.5% seen in cases who were controlled at metastatic site than those patients whose disease was persistent, and OS in these patients was 18.3% with p = 0.0003 [14].

7.1.3.1. Response rate in palliative radiotherapy

RT is effective in controlling metastatic deposits and in decreasing symptoms due to metastatic disease, with a response rate of 65.2% observed in various studies. Response rates varied with the type of metastases: 84.2% in soft tissue metastases and 63.2% in bone metastases [9].

7.1.4. Bone metastases

There are many established fears about the long-term impact of RT in children like growth disturbances or other skeletal abnormalities after irradiation of the musculoskeletal system. Radiotherapy is performed primarily in these patients with incurable or metastatic disease to relieve pain, definitely control a bone affected from metastases, and prevent pathologic fractures as well as spinal cord compression.

7.1.4.1. Radiotherapy doses for bone metastasis

Dose schedule ranges from 6 to 40 Gy [9]. Fraction regimens can be given from 1.8 to 4 Gy per fraction. The most commonly irradiated sites reported are the lower limb (34%) and spine (29%). An overall response rate of 65–77% is reported in various studies [9, 15, 16]. The Institut Curie, Paris, France, published a series of 23 children with neuroblastoma treated for 38 bony metastatic sites with doses ranging from 6 to 40 Gy. A trend toward improved response with higher doses was seen (**Tables 1** and **2**) [9].

7.1.5. Soft tissue metastasis

In soft tissue metastasis, mediastinum, supraclavicular, cervical, axillary, abdominal, and orbital sites are often involved.

7.1.5.1. Fractionation schedule

The optimal fractionation schedule is still an unresolved issue. Doses can be given as 20 Gy, with a range from 8 to 36 Gy [9]. Various fraction regimens, including 1.5–4 Gy per fraction, can be given in palliative setting. **Figure 6**(**a** and **b**) illustrates 3DCRT radiotherapy treatment of olfactory neuroblastoma with extension into maxillary sinus, ethmoid, orbital, and neck nodes metastasis.

Total dose	Response	p Value
<20 Gy	50%	0.088
≥20 Gy	81.2%	

Table 1. Response to radiotherapy according to dose delivered for bone metastases.

http://dx.doi.org/10.5772/intechopen.80892

Myc status	Good response	P value
Myc amplified	26%	0.70
Myc nonamplified	56.5%	

Table 2. Response to radiotherapy in 23 patients with bone metastasis according to MYCN status.

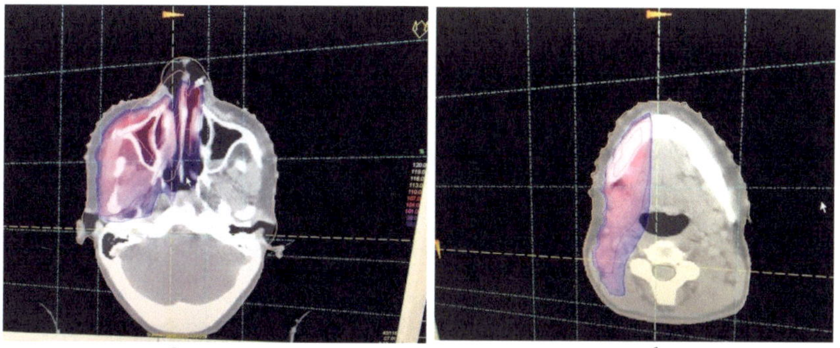

a b

Figure 6. (a and b) Isodose distribution for three-dimensional conformal radiotherapy (3DCRT) to the nose, orbit, right maxilla (**Figure 6a**), and ipsilateral neck (**Figure 6b**) with wedge for homogenous dose distribution. Blue represents 95% of the isodose line.

7.1.6. Radiotherapy for hepatomegaly (stage 4S)

Massive hepatomegaly causes respiratory distress. To prevent respiratory compromise, the liver is irradiated for symptomatic stage 4S disease. Doses can be given as 450–600 cGy in 2–4 fractions (450 cGy in 3 fractions, 1.5 Gy/fraction) [15]. Target volume can be treated by opposed lateral fields to avoid renal and ovarian exposure.

7.1.6.1. Radiotherapy borders

Radiotherapy box field should be gridded as anterior border 2 cm anterior to the liver, posterior border should be anterior to vertebral body, superior border 2 cm superior to the liver, and inferior border gridded at superior iliac crest to avoid ovarian exposure.

Although very rare but one study reported by Paulino showed survival of one patient stage 4S neuroblastoma with liver metastases who was alive 13 years after hepatic irradiation [15].

7.1.7. Intracranial metastasis of neuroblastoma (IMN)

CNS lesions are defined as leptomeningeal disease or metastatic deposits in the CNS parenchyma. Patients with radiographic evidence of bone involvement or with intracranial extension from the epidural, dura, or skull are not classified with CNS neuroblastoma. Patients who experience a disease recurrence incidence of leptomeningeal or CNS parenchymal disease range from 1 to 16%. The overall incidence rate for newly diagnosed patients is approximately 6.3%.

The incidence rate is >10% in cases who are treated with recent intensive chemotherapy and immunotherapy (N6 and N7, with which the expected cure rate is ≥38%) [17].

Symptoms of CNS involvement are listed in **Box 1**. Patients with neurologic symptoms should undergo physical examination and further neuroimaging with head/orbit CT and/or MRI scans. MIBG scan is not a reliable indicator of CNS disease as false-negative MIBG scans appear to be common for patients with CNS lesions. This is due to difficulty in discriminating CNS lesions from skull lesions without SPECT images.

- Headaches
- Nausea and emesis
- Seizures
- Motor weakness and paralysis
- Extremity or back pain
- Change in consciousness
- Cranial nerve symptoms
- Fever

Box 1. Clinical features of patients with central nervous system involvement.

7.1.7.1. Genetic associations in brain metastasis

Neuroblastoma tumors that exhibited MYCN amplification correlate with high-risk disease and poor prognosis as they developed CNS recurrence.

7.1.7.2. External beam radiotherapy in CNS metastasis

Prognosis is very poor even if multimodality treatment is given. Multiple brain metastasis can be treated by whole brain radiotherapy. In patients with oligometastasis in the brain, literature suggested that stereotactic radiosurgery (SRS) confers survival benefit [18].

7.1.8. Gamma knife radiosurgery

Nathan C. Rowland et al. described in literature two cases of the application of SRS to high-risk, recurrent IMN. Leksell Gamma Knife model 4C using Leksell GammaPlan 4C treatment planning software was used for radiotherapy. First patient 5-year-old male was treated case of neuroblastoma whose brain magnetic resonance imaging (MRI) showed a 4.0 × 3.3 cm mass in the right parietal lobe with a small amount of surrounding edema and midline shift. The dose prescription was 27.7 Gy to 47% isodose line. The 18 mm and 14 mm collimating helmets were used in 18 targets. Target volume was 29.2 cc and 100% target coverage was by prescribed isodose line. Conformity index of 1.34 was achieved. Radiological follow-up at 2 and 4 months revealed stable reduction of disease. Patient again received chemotherapy and 21.6 Gy dose given to craniospinal axis. Three-month post-radiotherapy intra-ommaya therapy along with complete surgical excision of mass was given. CNS relapse occurred 15 months after radiotherapy and patient gave up to disease. The second case was 2 ½-year-old male treated case of neuroblastoma which showed solid mass in the left parieto-occipital lobe measuring 1.8 × 1.0 cm with a small anterior cystic component. Stereotactic biopsy of the intracranial mass diagnosed the neuroblastoma. A dose of 14 Gy was prescribed to the 50% isodose line and this was administered to five targets. The 18 mm collimating helmet was utilized. Although radiological follow-up at 2 and 3 months showed reduction in size of initial parieto-occipital lesion, 4-month follow-up CT scan showed slight progression in size of metastatic lesion with vasogenic edema. Finally after CNS relapse which occurred 7 ½-month post-radiotherapy, patient died of disease. Stereotactic radiosurgery is an acceptable palliative method in the treatment of IMN [19].

7.1.9. Proton beam therapy

Proton beam therapy (PBT) is a good alternative to photon therapy for pediatric patients where post-radiotherapy side effects are concerned. PBT delivers radiation within a defined radiation track length, with virtually no dose beyond the intended target due to Bragg peak (**Figures 7** and **8**). Proton beam provides superior target volume coverage and greater dose reductions for normal tissues or organs at risk by a factor of 1.5–3.0 than photon beam. This is due to additional benefit of no exit dose and low entrance dose [20]. Proton beam is considered safe for neuroblastoma pediatric patients with minimal risk of secondary cancer which is slightly higher with intensity-modulated radiotherapy (IMRT) due to peripheral doses. As discussed earlier, doses in path of proton beam are very minimal. There is homogenous dose distribution with large volume of neuroblastoma with less number of ports than IMRT [21].

7.1.10. Procedure

Before treatment, CT images for PBT planning are to be obtained at intervals of 2–5 mm in the treatment position. The interval depends on the age, height, and treatment site of the patient. The gross tumor volume (GTV) is defined as the tumor volume before PBT for a recurrent tumor. The clinical target volume (CTV) is defined as the GTV plus a 1.5 cm margin, and the PTV is defined as the CTV plus a 0.5–0.7 cm margin. Toxicity and treatment effect should be

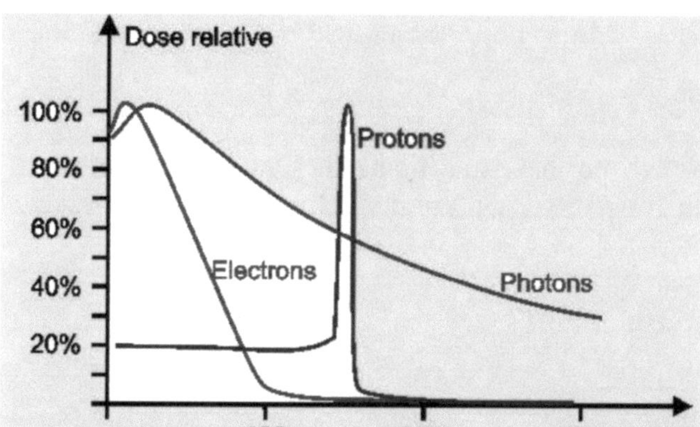

Figure 7. Bragg peak of proton is very sharp.

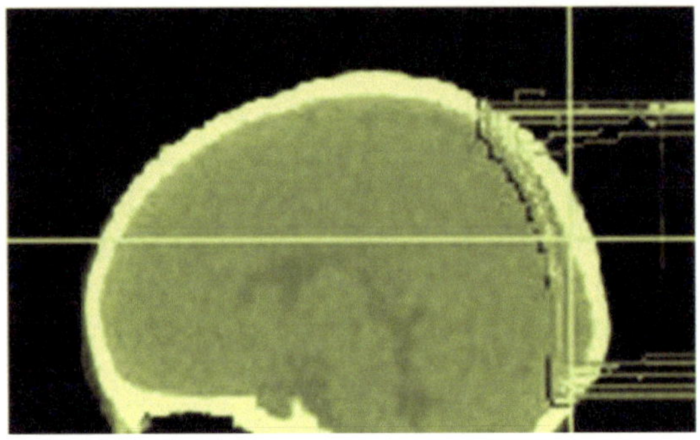

Figure 8. Sharp reduction of dose to brain tissue by proton irradiation.

balanced in determining the CTV. Sedatives can be administered for planning CT and treatment in pediatric patients.

The photon equivalent dose (GyE) = [Physical Dose (Gy)] × [Relative Biological Effectiveness of the proton beam]. This is assigned with a value of 1.1.

The PBT doses usually administered from 19.8 to 45.5 GyE (median: 30.6 GyE) [21].

8. Summary

- Radiotherapy can be considered to primary site or residual MIBG-positive metastatic site.
- Palliative radiotherapy is given on individual case to case basis.
- Abdominal and pelvic sites are commonly treated with AP fields or 3DCRT techniques. Plans should be with the use of multileaf collimators (MLC).
- Dose: 21.6 Gy in 12 fractions (1.8 Gy per fraction).
- GTV (primary tumor volume) = pre-surgery CT/MIGB scans.
- Bone metastatic site = volume positive on MIGB/bone scan after induction chemo.
- PTV = GTV + 2 cm.

Dose constraints for liver normal parenchyma are V9 Gy <50% and V18 Gy <25%; contralateral kidney constraints are V8 Gy <50% and V12 Gy <20%.

Acknowledgements

We gratefully acknowledge Dr Vijay Dhasmana, the Honorable Vice Chancellor of Swami Rama Himalayan University, Dehradun, India, for his support and for providing the necessary facilities, and we are also thankful to Prof. (Dr) Sunil Saini, Director of Cancer Research Institute, Chairman Research Committee, Himalayan Institute of Medical Sciences, Swami Rama Himalayan University and Prof. (Dr) Mushtaq Ahmad, Dean, Himalayan Institute of Medical Sciences, Swami Rama Himalayan University, Dehradun.

Conflict of interest

We have no conflict of interest with anybody working in the area and among the authors in the book chapter.

Author details

Meenu Gupta[1*] and Anupam Dhasmana[2]

*Address all correspondence to: meenugupta.786@rediffmail.com

1 Department of Radiation Oncology, Cancer Research Institute, Swami Rama Himalayan University, Dehradun, India

2 Department of Biosciences, Swami Rama Himalayan University, Dehradun, India

References

[1] Kelly EB. Neuroblastoma: Cancer in Very Young Children. The 101 Most Unusual Diseases and Disorders. Vol. 1; Nov 2015. p. 253

[2] Goodman MT, Gurney JG, Smith MA, Olshan AF. Sympathetic nervous system tumors. In: Cancer Incidence and Survival among Children and Adolescents: United States SEER Program, 1975-1995. Bethesda, MD: National Cancer Institute; 1999. p. 35

[3] Maris JM. Recent advances in neuroblastoma. The New England Journal of Medicine. 2010;**10**:2202-2211. DOI: 10.1056/NEJMra0804577

[4] Chu CM, Rasalkar DD, Hu YJ, Cheng FWT, Li CK, Chu WCW. Clinical presentations and imaging findings of neuroblastoma beyond abdominal mass and a review of imaging algorithm. The British Journal of Radiology. Jan 2011;**84**(997):81-91

[5] Hadley L, Lakhoo K. Book Chapter 105. Neuroblastomas: Pediatric Surgery: A Comprehensive Text for Africa. Vol. 2; 2011. p. 610

[6] DuBois SG, Kalika Y, Lukens JN, Brodeur GM, Seeger RC, Atkinson JB, et al. Metastatic sites in stage IV and IVS neuroblastoma correlate with age, tumor biology, and survival. Journal of Pediatric Hematology/Oncology. May-Jun 1999;**21**(3):181-189

[7] Pepper WA. A study of congenital sarcoma of the liver and suprarenal with report of a case. American Journal of the Medical Sciences; (1827-1924). 1901;**121**:287-299

[8] Malis J. Clinical Presentation of Neuroblastoma. Chapter 1, IntechOpen. 2013:3-12. http://dx.doi.org/10.5772/55921

[9] Caussa L, Hijal T, Michon J, Helfre S. Role of palliative radiotherapy in the management of metastatic pediatric neuroblastoma: A retrospective single-institution study. International Journal of Radiation Oncology, Biology, Physics. Jan 1, 2011;**79**(1):214-219

[10] Monclair T, Brodeur GM, Ambros PF, Brisse HJ, Cecchetto G, Holmes K, et al. The international neuroblastoma risk group (INRG) staging system: An INRG task force report. Journal of Clinical Oncology. 10 Jan 2009;**27**(2):298-303

[11] Katsanis E. Whitesell L. Neuroblastoma. Elsevier BV; 2002

[12] Kandula S, Prabhu RS, Nanda R, Switchenko JM, Cash T, Qayed M, et al. Outcomes after radiation therapy to metastatic sites in patients with stage 4 neuroblastoma. Journal of Pediatric Hematology/Oncology. Apr 2015; **37**(3):175-180

[13] Mazloom A, Louis CU, Nuchtern J, Kim E, Russell H, Allen- Rhoades W, et al. Radiation therapy to the primary and Postinduction chemotherapy MIBG-avid sites in high-risk neuroblastoma. International Journal of Radiation Oncology, Biology, Physics. Nov 15 2014; **90**(4):858-862

[14] Casey DL, Kushner B, Cheung NK, Modak S, Wolden SL. Radiation therapy to sites of metastatic disease as part of consolidation in high-risk neuroblastoma: Can long-term control be achieved? International Journal of Radiation Oncology, Biology, Physics. Oct 1, 2017;**99**(Issue 2, Supplement):E565

[15] Paulino AC. Palliative radiotherapy in children with neuroblastoma. Pediatric Hematology and Oncology. Mar 2003;**20**(2):111-117

[16] Halperin EC, Cox EB. Radiation therapy in the management of neuroblastoma: The Duke University medical Center experience 1967-1984. International Journal of Radiation Oncology, Biology, Physics. 1986;**12**:1829-1837

[17] Kramer K, Kushner B, Heller G, Nai-Kong VC. Neuroblastoma metastatic to the central nervous system. Cancer. April 15, 2001;**91**(8):1510-1518

[18] Niibe Y, Karasawa K, Nakamura O, Shinoura N, Okamoto K, Yamada R, et al. Survival benefit of stereotactic radiosurgery for metastatic brain tumors in patients with controlled primary lesions and no other distant metastases. Anticancer Research. 2003 Sep-Oct;**23**(5b):4157-4159

[19] Rowland NC, Andrews J, Patel D, LaBorde DV, Nowlan A, George B, et al. Intracranial metastatic neuroblastoma treated with gamma knife stereotactic radiosurgery: Report of two novel cases. Case Reports in Neurological Medicine. 2012;**2012**:690548

[20] Main C, Dandapani M, Pritchard MG, Wilson J. The effectiveness and safety of proton beam radiation therapy in children with malignant central nervous system (CNS) tumours. Systematic Reviews. Dec 2016;**5**(1):1-7

[21] Oshiro Y, Mizumoto M, Okumura T, Sugahara S, Fukushima T, Ishikawa H, et al. Clinical results of proton beam therapy for advanced neuroblastoma. Radiation Oncology. 2013;**8**:142

Chapter 3

Effect of Low Dose (Diagnostic X-Rays) on Peripheral White Blood Cells Count in Guinea Pigs (*Cavia porcellus*)

Geofery Luntsi, Victory S. Daniel,
Chigozie I. Nwobi and Bura T. Paul

Additional information is available at the end of the chapter

http://dx.doi.org/10.5772/intechopen.80417

Abstract

Exposure to ionizing radiation is known to affect some hematological parameters of biological sample. This study was aimed at evaluating the effect of ionizing radiation within the diagnostic range on some hematological parameters in guinea pigs. Thirty six (36) apparently healthy adult guinea pigs of both sexes weighing between 700 and 1200 g were used. The guinea pigs were categorized in to three groups, 12 per group; group A (control), group B, and C were exposed to X-rays within the diagnostic range, using 70 kV and 12.5mAs; using X-ray machine MS-185, serial no. 0904 GE at a source to skin distance (SSD) of 90 cm. Blood samples were collected from all the guinea pigs at intervals of 1, 24, 72, 168 and 336 hours post-irradiation, and subjected to standard hematological analysis. A continuous decline in the mean total white blood cell count and mean lymphocyte, monocyte, neutrophil and eosinophil count after 1 hour in both groups was observed, and more pronounced after 24 hours post-irradiation. However, stability was observed 72 hours post-irradiation in both groups. In conclusion, a depleting effect of low dose ionizing radiation on white blood cell count was found, with appreciable recovery occurring after 72 hours onward.

Keywords: white blood cell, irradiation, guinea pigs, ionizing radiation, hematological parameters

1. Introduction

Radiation is a wave or particle traveling through space which can transmit all or parts of its energy on contact with matter [1]. It could be ionizing and non-ionizing in nature [2].

Ionizing radiation is a very high-energy form of electro-magnetic radiation which has the energetic potential to break apart electrically neutral atoms resulting in the production of negative and/or positive ions [3]. Non ionizing radiation is relatively a low-energy radiation that does not have sufficient energy to ionize atoms or molecules [4]. Although considered less dangerous than ionizing radiation, over exposure to non-ionizing radiation can also be hazardous [4].

Exposure to radiation results in a deposition of energy in tissues that can damage cellular structures including DNA [5]. The degree of the damage due to the radiation depends on the type of radiation, energy of the radiation, intensity of the radiation, and exposure time [6]. Depending on the duration of exposure, the area exposed and the dose received, radiation exposure in the immediate aftermath could lead to a myriad of deleterious effects including acute radiation syndrome [7]. Acute radiation syndrome (ARS) includes hematopoietic syndrome, gastro-intestinal syndrome and cardiovascular/central nervous system syndrome among others. Hematopoietic syndrome may occur after exposure to significant radiation dose and all blood components may be affected adversely [8]. Blood being a vital special circulating tissue composed of cells suspended in a fluid (plasma) with a major function of maintaining homeostasis [9] may experience decline in cell count on exposure to ionizing radiation leading to drop in circulating blood cells which is detrimental to the health of the individual [10]. The white blood cells also called leukocytes are the mobile units of the body's protective system [11]. Decrease in the WBC count leaves the individual at risk of infection. Low WBC count is known as leucopenia.

Guinea pigs are rodents of the family Caviidae and the genus *Cavia* which are mostly kept as pets and also used as laboratory animals for biomedical experiments. *Cavia porcellus* are small stout-bodied short-eared tailless domesticated rodent of South American origin [12]. They are not related to swine neither are they from Guinea Republic [13]. They are used for meat, local medicine and play important roles in religious and cultural ceremonies especially in South America [13]. Guinea pigs are used for biomedical research because they are biologically similar to humans as they share more than 90% DNA with humans, diseases that affect humans are also likely to affect them and they have shorter life span making it possible for them to be studied throughout their life time and they also are easy to handle [14].

Ionizing radiation is widely used in the medical field for both diagnostic and therapeutic purposes in form of X-rays, gamma rays, and particles (α-particle, β-particle, protons and neutrons) radiations [15].

The biological effect of ionizing radiation arises from the deposition energy in the tissues which can cause changes in the chemical composition of the cell. The energy of the ionizing radiation is significantly greater than the bond energies of many molecules and can cause homolytic bond scission and generation of secondary electrons [6]. Ionizing radiation is thus seen to affect biological tissues by directly dissociating molecules following their excitation and ionization, or indirectly by the production of free radicals and hydrogen peroxide in the water of the body fluids [15], and the severity of the effect increases with dose and dose rate [16].

Although the use of ionizing radiation involves a certain level of risk, its use in medicine results in such numerous benefits that if judiciously employed, the benefits greatly exceed the risk to the individual [17]. The hematopoietic system is highly sensitive to radiation, and peripheral blood examination may serve as a biological indicator of such damage that may occur even at very low doses of ionizing radiations like X-rays or gamma rays [15, 18]. Peripheral blood examination may serve as a screening test for various hematological as well as non-hematological disease states [18]. Radiographic imaging is extremely valuable as a diagnostic and therapeutic tool in medicine, but ionizing radiation also carries well-known potential risk [19]. It is generally known that exposure to high energy ionizing radiation like X-ray is known to have effect on rapidly dividing cells of the bone marrow, blood, and mucosal linings. Often, patients are required to undergo repeated exposure which usually increases the risk of damage by ionizing radiation damage on the hematopoietic system [17]. There is no known safe period for the patient to undergo repeated exposure with less or minimal risk to ionizing radiation. There is dearth of information on the studies bordering on the Effect of Ionizing Radiation on White blood cells within the locality of study.

The white blood cells fights off infections and defends the body against invasion by foreign organisms and to produce or at least transport and distribute antibodies in immune response [20]. Decrease in the white blood cell count leaves the individual at risk of infection. Low white blood cell count is known as leucopenia [11]. It has been observed that there is always a slight decrease in the total white cells count after the first few days of exposure to ionizing radiation; hence, white blood cells count may be a reliable indicator of degree of exposure [17]. Irradiating animal models to a single whole-body dose of ionizing radiation result in complex sets of symptoms whose onset, nature, and severity are functions of both total radiation dose and radiation quality which are classified into three syndromes: the hematopoietic syndrome, the gastrointestinal syndrome, and the central nervous syndrome. The hematopoietic syndrome occurs at very low radiation doses and is manifested by depletion of hematopoietic stem cells and ultimately by depletion of matured hematopoietic and immune cells [21]. This study was aimed at observing the changes that may occur on the white blood cells counts after exposure to low dose ionizing radiation (X-rays) within the diagnostic range, using guinea pigs (*Cavia porcellus*) as animal sample.

1.1. White blood cells (WBC)

The white blood cells also known as leukocytes make up approximately 1% of the total volume of the cells in the blood [22]. The WBCs are primarily involved in the immune response and defense of the body. The WBCs differs from the red blood cells (RBC) as they do not have nuclei and do not contain hemoglobin [11, 23]. The WBCs are formed in the bone marrow and lymph tissue which are then transported to different locations of the body where it is needed. The number of WBCs in the blood is often an indicator of disease, significant increase in the number is known as leukocytosis and significant decrease is called leucopenia [22].

1.1.1. Classification

There five types of WBCs which are classified into two major groups: granular and agranular WBC [24]. The granular WBCs are: neutrophils, eosinophils, and basophils. The granulocytes are characterized by a lobed nucleus and granular inclusions in the cytoplasm. Granulocytes are typically first-responders during injury or infection [11, 23]. The agranular WBCs are; lymphocytes and monocytes. The lymphocytes include B and T cells and are responsible for adaptive immune response. The monocytes differentiate into macrophages and dendritic cells, which in turn respond to infection or injury [11, 23].

1.2. Biologic effect of radiation

Soon after the discovery X-rays and radioactivity it became evident that ionizing radiation could cause damage to cells and tissues [16]. For risk estimation, scientists presently rely on molecular, cellular and animal experiments. The immediate effect of ionizing radiation is directly cellular damage through ionization, excitations and indirect damage by formation of radicals that initiate chemical reactions occur within a very short period following exposure. Subsequently these effects induce changes at the level of molecules (e.g., DNA) [25]. The interaction of radiation and the tissue is governed by the energy and mass of the incident radiation (alpha, beta particle, gamma ray or X-ray) and the properties of the tissue [26]. If the damage is not or not correctly repaired, cell, tissues and finally the whole organism may be affected. Above small doses (few grays), cell death is the dominant effect, which may cause severe damage to organs and tissues [25]. Other effects occur long after the exposure and involve the risk of developing radiation-induced cancer and hereditary disease in the offspring of following generations of the exposed persons.

1.2.1. Deterministic effect

Radiation kills cells at high exposures. Low numbers of dead cells will usually be replaced through cell division in a tissue or organ, but if the numbers of killed cells is too large, harm occurs to the tissue or organ [25]. The deterministic effects occur at high dose level, in which below the dose the effect will not be observed. The severity of the effect increases with dose and dose rate [25]. Fortunately deterministic effects are perceived at relatively high doses are there hardly observed in diagnostic radiology because of the low doses used. Exceptions are incident with deterministic radiation induced skin injury after prolong fluoroscopy-guided procedures. Most deterministic effects come early to expression even though some can occur later [25].

1.2.2. Stochastic effects

Radiation-induced malignancies and heritable effect are referred to as stochastic effects. These effects do not have threshold, this implies that there is finite probability they can occur after exposure to very low doses of radiation [25]. For stochastic effects not the severity but the likelihood of occurrence of the effect depend on the dose, therefore the probability of occurrence depends on with increasing dose [2]. Theoretically a single ionization track has the potential to result in a detrimental stochastic effect.

http://dx.doi.org/10.5772/intechopen.80417

2. Materials and methods

Institutional approval to conduct the study was obtained from the committee on ethics of the Veterinary teaching hospital, University of Maiduguri (VTH). Thirty six (36) guinea pigs were obtained and kept at the large animal clinic of the VTH, Faculty of Veterinary Medicine University of Maiduguri, under good ventilation and adequate light. The guinea pigs were fed with standard commercial prepared diet (pelletized feed) and vegetables (such as cabbage and carrots) and given free access to clean drinking water. The guinea pigs were kept in this condition for 14 days in order to acclimatize before starting the experiment [27]. The guinea pigs were routinely screened for ectoparasites, endoparasites, and hemoparasites using standard methods by a veterinary doctor, and randomly divided into three groups, 12 guinea pigs per group. Group A served as the control group, group B and group C were exposed to low dose X-rays at a dose that is within the diagnostic range, using factors for chest X-ray of an adult patient in the study center (70 kV and 12.5mAs) using X-ray machine MS-185, serial no. 0904 GE, on which quality assurance check was routinely performed by a medical physicist with over 8 years experience, at a source to skin distance (SSD) of 90 cm. The guinea pigs in each group were irradiated together using a vertical central ray on a horizontal table top (couch) within the same cage, with the radiation properly collimated to include all the guinea pigs. Group C was irradiated twice with the same exposure factors 5 minutes after the first exposure.

2.1. Recruitment of subjects

A total of 36 adult guinea pigs of both sexes, weighing between 700 and 1200 g, were used for the study although 50 was obtained in case of accidental death, straying away, and some may be sickly.

2.1.1. Inclusion criteria

Thirty six apparently healthy adult guinea pigs of both sexes were used for the study.

2.1.2. Exclusion criteria

Apparently (physically) unhealthy and diseased guinea pigs were not selected for this study.

2.2. Irradiation procedures

Group A served as the control group. Group B and group C were irradiated with X-ray dose of about 70 kV and 12.5mAs which is within the diagnostic range from X-ray machine MS-185, serial no. 0904 GE at focus to film distance (FFD) of 90 cm. The guinea pigs in each group were irradiated together with a vertical central ray on a horizontal table top. Group C were irradiated again with the same exposure factors 5 minutes after the first exposure.

2.3. Blood sample collection

Blood sample from each guinea pig was collected into EDTA bottle from direct cardiac puncture with a 2 ml syringe and appropriately labeled. The blood samples were collected at the intervals of 1, 24, 72, 163 and 336 hours post irradiation of the experimental groups. Blood samples were also collected during same time interval from the control group. No same syringe was used to collect blood sample more than once. The blood cell count was done by a veterinary doctor with over 10 years experience in veterinary parasitology at the veterinary teaching hospital University of Maiduguri, who performed the procedure alone to avoid inter-observer error. Hemocytometric method was used to count the white blood cells using Neubauer counting chamber. This method was used due to availability and convenience, as the automatic analyzer was not readily available at the time of analysis.

2.4. Hematological examination

All blood samples collected were subjected to standard hematological procedures to determine PCV, Hb, WBC and differential WBC count.

White blood cell count: Bulk dilution of the white blood cell count was employed. 0.02 ml of well mixed EDTA anticoagulant blood was pipetted into 0.38 ml of Turks solution contained in Khan tube and mixed. A clean cover slip was put in place on the improved Neubauer counter. Using a capillary tube held at an angle of 45° to the counting chamber, the diluted blood sample was carefully discharged into the counting chamber. The chamber was then placed in a petri dish and left undisturbed for 2 minutes, allowing the cells to settle. The underside of the chamber was dried and placed on a microscope and examined with 10× objective. The cells in the four large corners of each chamber were counted, including cells on the lines of two sides of the large squares. The number of white cells (per liter of blood) was recorded using a correction factor of 10×.

Differential white blood count: Longitudinal method of differential white blood cell count was adopted. A drop of blood was placed on a clean dry glass slide and a thin film was made. The film was dried in the air, fixed and stained by flooding with Leishman stain and allowed to stand for 30 minutes. Then the excess stain was washed off and allowed to dry in the air. A drop of immersion oil was placed on the film and covered with a clean dry cover slip. The film was viewed under 100× objective of the microscope. The differential white cells seen in each field was counted using the automated differential cell counter and recorded appropriately. Thus, the observed number of WBC indices in response to irradiation is used as an indicator of exposure [28].

2.5. Statistical analysis

The mean values of hematological parameters of control, single and double exposure groups were determined using one way analysis variance (ANOVA). P-values <0.05 was considered significant and the mean ± SE for hematological parameters were presented using descriptive statistics.

3. Results

The mean ± SE of white blood cell count values of guinea pigs for the control group is shown in **Table 1**, while **Tables 2** and **3** shows the mean ± SE of white blood cell count values of guinea pigs following single and double exposures to X-rays within diagnostic range respectively.

http://dx.doi.org/10.5772/intechopen.80417

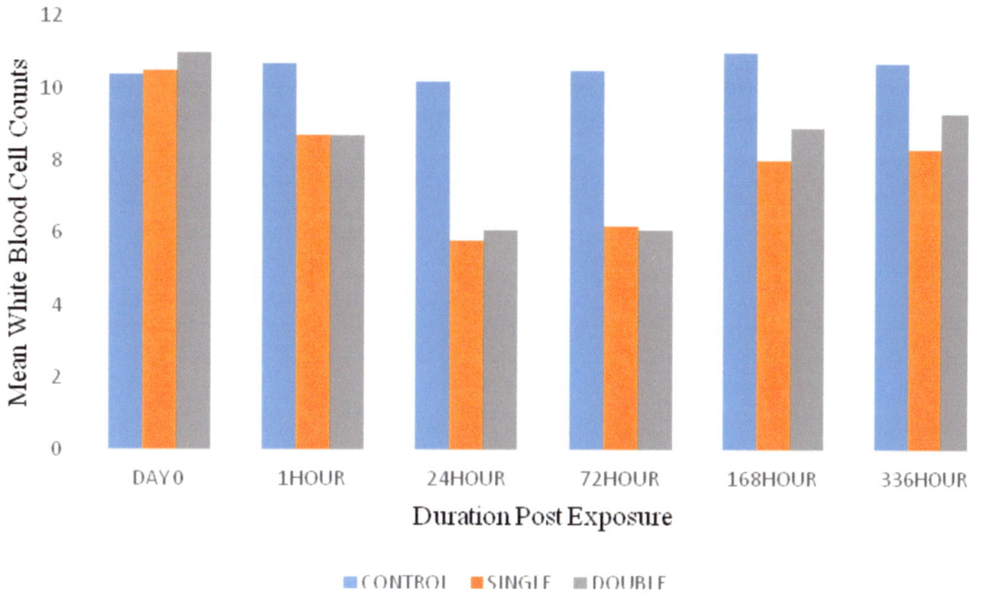

Figure 1. Mean total white blood cell counts of guinea pigs (*Cavia porcellus*) following single and double exposure to radiation.

There was an observed decline in the mean total white blood cell count of guinea pigs after 1hour post exposure to single and double exposures **Figure 1**. This was more pronounced after 24 and 72 post exposures. Marked recovery of WBC was noticed after 168 and 336 hours post exposure in both single and double exposure groups.

A decline in the mean total white blood cell count of guinea pigs was observed at 1 hour after single exposure, and was more pronounced 24 hours post irradiation. However, recovery of WBC commenced at 72 hours post exposure after single exposure. This was found to be significant ($p < 0.05$) as shown in **Table 2**.

Parameters	**Control group**					
	Mean ± SE					
	0	**1 hr**	**24 hr**	**72 hr**	**168 hr**	**336 hr**
WBC	10.4 ± 0.8	10.7[a] ± 0.8	10.2[a] ± 0.6	10.5[a] ± 0.5	11.0 ± 1.0	10.7 ± 0.4
PCV	40 ± 1.8	40.3 ± 1.8	40.5 ± 1.5	38.5 ± 1.9	38.8 ± 0.8	39.3 ± 1.4
HB	13.0 ± 0.4	12.8 ± 0.9	12.8 ± 0.9	12.5 ± 0.3	12.3 ± 0.3	12.3 ± 0.3
Monocyte	0.4 ± 0.2	0.3[a] ± 0.1	0.3[a] ± 0.1	0.4[a] ± 0.1	0.4[a] ± 0.1	0.3[a] ± 0.1
Lymphocyte	4.7 ± 0.7	5.2 ± 0.6	4.9[a] ± 0.5	4.7[a] ± 0.6	4.9 ± 0.7	4.7 ± 0.4
Neutrophil	5.5 ± 0.7	5.7 ± 0.8	5.2[a] ± 0.6	5.2[a] ± 0.5	5.7 ± 0.8	5.5 ± 0.3
Eosinophil	0.4 ± 0.07	0.4[a] ± 0.1	0.4[a] ± 0.1	0.4[a] ± 0.1	0.4[a] ± 0.1	0.4[a] ± 0.06
Basophil	00	00	00	00	00	00

Key: a—not significant and b—significant ($P > 0.05$).

Table 1. Effects of low radiation dose exposures on hematological parameters of guinea pigs (*Cavia porcellus*).

Parameters	Single exposure					
	Mean ± SE					
	0	1 hr	24 hr	72 hr	168 hr	336 hr
WBC	10.5 ± 0.5	8.7^{a} ± 0.9	5.8^{b} ± 0.8	6.2^{b} ± 0.9	8.0 ± 1.1	8.3 ± 1.2
PCV	40 ± 1.1	35.3 ± 3.3	28.9 ± 4.0	27.4 ± 3.8	30.3 ± 4.2	31.6 ± 4.4
HB	12.6 ± 0.3	11.2 ± 1.1	9.0 ± 1.2	8.8 ± 1.2	9.3 ± 1.3	10.0 ± 1.4
Monocyte	0.3 ± 0.0	0.2 ± 0.0	0.2^{b} ± 0.0	0.1^{b} ± 0.0	0.2^{b} ± 0.0	0.2^{b} ± 0.0
Lymphocyte	3.7 ± 0.5	4.0 ± 0.5	2.5^{b} ± 1.3	2.6^{b} ± 0.4	3.2 ± 0.5	3.7 ± 0.6
Neutrophil	5.5 ± 0.4	4.3 ± 0.6	2.7^{b} ± 0.0	2.8^{b} ± 0.5	3.8 ± 0.6	4.2 ± 0.6
Eosinophil	0.4 ± 0.03	0.2^{b} ± 0.0	0.1^{b} ± 0.4	0.1^{b} ± 0.0	0.1^{b} ± 0.0	0.1^{b} ± 0.0
Basophil	00	00	00	00	00	00

Key: a—not significant and b—significant ($P > 0.05$).

Table 2. Effects of low radiation dose single exposures on hematological parameters of guinea pigs (*Cavia porcellus*).

Parameters	Double exposure					
	Mean ± SE					
	0	1 hr	24 hr	72 hr	168 hr	336 hr
WBC	11.0 ± 0.4	8.7^{a} ± 0.4	6.1^{b} ± 0.2	6.1^{b} ± 0.2	8.9 ± 0.3	9.3 ± 0.3
PCV	41 ± 1.1	38.5 ± 1.0	31.8 ± 1.0	34.3 ± 1.3	38.0 ± 1.2	39.5 ± 0.9
HB	13.1 ± 0.3	11.9 ± 0.5	9.8 ± 0.4	10.8 ± 0.3	11.7 ± 0.3	12.3 ± 0.2
Monocyte	0.3 ± 0.1	0.1^{b} ± 0.0	0.1^{b} ± 0.0	0.1^{b} ± 0.0	0.2^{b} ± 0.0	0.2 ± 0.0
Lymphocyte	6.0 ± 0.4	4.4 ± 0.3	2.9^{b} ± 0.2	2.7^{a} ± 0.1	3.8 ± 0.2	4.7 ± 0.3
Neutrophil	5.8 ± 0.3	4.3 ± 0.3	2.8^{b} ± 0.1	2.7^{b} ± 0.1	4.1 ± 0.2	4.5 ± 0.2
Eosinophil	0.5 ± 0.04	0.2^{b} ± 0.0	0.1^{b} ± 0.0	1.1^{b} ± 0.0	0.2^{b} ± 0.0	0.2^{b} ± 0.0
Basophil	00	00	00	00	00	00

Key: a—not significant and b—significant ($P > 0.05$).

Table 3. Effects of low radiation dose double exposures on hematological parameters of guinea pigs (*Cavia porcellus*).

There was also a decline in the mean total white blood cell count of guinea pigs at 1 hour after double exposure. This was more pronounced after 24 hours post irradiation. The recovery of WBC was observed 72 hours after double exposure group. This was found to be significant ($p < 0.05$) as seen in **Table 3**.

A decline in the mean absolute monocyte count at 1 hour post irradiation was also noted. This decrease was more pronounced at 24–72 hours post irradiation. However, recovery of monocytes was evident at 168–336 hours post irradiation in both single and double exposure groups, with no significant difference ($p < 0.05$) as shown in **Tables 2** and **3**.

http://dx.doi.org/10.5772/intechopen.80417

There was also a slight decline in the mean absolute lymphocyte count of guinea pigs at 1 hour in the exposure groups, which was more pronounced 24–72 hours post irradiation in both exposure groups. However, recovery of the mean absolute lymphocyte count was evident at 168 and 336 hours post irradiation as seen in **Tables 2** and **3**.

There was an observed decline in the mean absolute eosinophil count of guinea pigs in both single and double exposure group at 1 hour following single and double exposure to irradiation. This decrease was sustained and was more pronounced 24–72 hours post irradiation. However, there was slight recovery of the mean absolute eosinophil count at 168–336 hours post irradiation, as shown in **Tables 2** and **3**.

An observed decline in the mean absolute neutrophil count of guinea pigs at 1 hour in both single and double exposure groups, and became more pronounced 24 hours after irradiation. However, slight recovery of mean absolute neutrophil count in guinea pigs was observed at 168–336 hours post irradiation as seen in **Tables 2** and **3**.

4. Discussion

A decrease in total white blood cell count; lymphocytes, monocytes, neutrophils, eosinophil was observed; however, basophils were not seen. This probably could be because basophils naturally are rarely encountered granulocytes in the peripheral blood, therefore, it is not unusual for basophils to be absent [29]. Previous studies have reported similar findings [6, 17, 30]. The observed decline in the white blood cell counts could be attributed to high radio-sensitivity of hematopoietic tissues [6, 31]. The results are consistent with the previous findings that irradiation induces leucopenia and reduces lymphocytes, neutrophils and monocytes count [32, 33]. However, the recovery was evident 72 hours post irradiation and onward, even though the recovery and repair took longer time than the damage [17, 33]. This could be due to the fact that the recovery might be as a result of the repair at the cellular level where sub-lethally damaged cells recover their viability and proliferation of undamaged cell elements [17, 30]. The effect on the double exposure group was severe, which proves the fact that severity of damage increases with increase in dose or exposure [16, 30, 33].

5. Conclusion

This study found a depleting effect of low dose ionizing radiation on the white blood cell counts of guinea pigs (*Cavia porcellus*). This was found to be more pronounced with repeated exposures. However, recovery occurred from 3 days (72 hours) post irradiation onwards. Thus, a proposed interval of 3–14 days (72–336 hours) before repeating an exposure is recommended for subjects that may require a series of follow up and repeat radiographic examinations.

Acknowledgements

We appreciate the entire staff of the Department of Medical Radiography, and the staff of Veterinary Teaching Hospital, University of Maiduguri, for their immense contribution throughout the course of this project.

Conflict of interest

Nil.

Author details

Geofery Luntsi[1]*, Victory S. Daniel[1], Chigozie I. Nwobi[1] and Bura T. Paul[2]

*Address all correspondence to: geostuffy@unimaid.edu.ng

1 Department of Medical Radiography, College of Medical Sciences, University of Maiduguri, Borno State, Nigeria

2 Veterinary Teaching Hospital, University of Maiduguri, Borno State, Nigeria

References

[1] Amon T, Oberhumer H. Basic Scientific Information and Facts about Radiation. 2007. Available at: http://www.radiationgenes.net/0nzm/CellTissueHumanBody6/o3aaklm/Radiation.htm [Accessed: 4 January 2016]

[2] Bushberg T, Seibert A, Laidholdt E, Boon J. The Essential Physic of Medical Imaging. 2nd ed. USA: Lippicott William and Wilkins; 2002

[3] Saia DA. Radiography PREP. 5th ed. USA: McGraw Hill; 2009

[4] Luntsi G, Muftaudeen B, Charles UE, Nwobi IC, Abubakar A, et al. Evaluation of the knowledge and awareness of non-ionizing radiation among final year students of college of medical sciences university of Maiduguri. International Research Journal of Pure and Allied Physics. 2015;**3**(3):8-14

[5] Koenig K, Goans R, Hatchett R, et al. Medical treatment of radiological casualties: Current concepts. Annals of Emergency Medicine. 2005;**45**:643-652

[6] Abojassim AA, Jaffat HS, Hassan AB. Effect of gamma radiation on some Haematological parameters in female rats. International Scientific Journal: Theoretical and Applied Science. 2015;**25**(5):101-109. Available at: http://www.T-science.org

[7] Citrin D, Cotrim A, Hyodo F, Baum B, Krishna M, Mitchell J. Radioprotectors and mitigators of radiation-induced normal tissue injury. The Oncologist. 2010;**15**:360-371. Available at: http://www.ro-journal.com

[8] Radiation Emergency Medical Management (REMM) [online]. 2014. Available at: www.remm.nlm.gov/ars_wbd.htm. [Accessed: 8 August 2014]

[9] NseAbasi N. E, Mary E. W, Uduak A, Edem E. A. O. Haematological Parameters and Factors Affecting Their Values. Agricultural Science. 2014;**2**(1):37-47

[10] Fliedner T, Friesecke I, Graessle D, Paulsen C, Weiss M. Hematopoietic cell renewal as the limiting factor in low-level radiation exposure: Diagnostic implications and therapeutic options. Military Medicine. 2002;**167**:46-48

[11] Guyton, Authur C, Hall, John E. Text Book of Medical Physiology. 11th ed. Philadelphia: Saunders Company; 2006

[12] Royal Society for the Prevention of Cruelty to Animals (RSPCA). Guinea Pigs: Good Practice for Housing and Care. 3rd ed. London: Research Animal Department; 2011

[13] Denise N. The Guinea Pig. ANZCCART News. 1994;**7**(3):1-8

[14] California Biomedical Research Association. 2007. Why animals are necessary in biomedical research: CBRA Fact Sheet. www.ca-biomed.org. [Accessed: 15 January 2016]

[15] Saman S, Nasir M, Muhammad NC, Shaharyar S, Nauman A. Assessment of impacts of hematological parameters of chronic ionizing radiation exposed workers in hospitals. FUUAST Journal of Biology. 2014;**4**(2):135-146

[16] Geleijns J, Broerse J, Brugmans M. Health effects of radiation exposure in diagnostic radiology. European Radiology Supplements. 2004;**14**:19-27

[17] Ikamaise VC, Chiaganam NO, Dike EU, Ndem BN. Post irradiation cell recovery in rats. Nigeria Journal of Medical Imaging and Radiation Therapy. 2007;**1**(1):18-23

[18] Mazin RM, Shaymaa MA, Najm AD, Mustafa GT, Shatha AJ, Abdulkareem HA. Effects of radiation on the hematological parameters in X-ray technicians: A case-control study. Journal of Pioneering Medical Sciences. 2014;**4**(2):85-88

[19] Salerno S, Paola M, Andrea M, Paolo T, Domenica M, et al. Radiation risks knowledge in residents and fellows in paediatrics: A questionnaire survey. Italian Journal of Paediatrics. 2015;**41**:21. DOI: 10.1186/s13052-015-0130-x

[20] NseAbasi NE, Mary EW, Uduak A, Edem EA. Haematological parameters and factors affecting their values. Agricultural Science Journal. 2014;**2**(1):37-47

[21] Singh A, Kumar R, Nivedita, Singh JK, Tanuja. Radioprotective effect of *Eclipta alba* (L.) against radiation induced haematological changes in Swiss albino mice. Journal of Natural Products. 2011;**4**:177-183

[22] Bruce A, Alexander J, Julian L, Martin R, Keith R, Peter W. Leukocyte Functions and Percentage Break Down. Molecular Biology of the Cell. 4th ed. New York: Garland Science; 2002

[23] Chiu S, Bharat A. Role of monocytes and macrophages in regulating immune response following lung transplantation. Current Opinion in Organ Transplantation. 2016;**21**(3): 239-245

[24] Haugland GT, Jordal A-EO, Wergeland HI. Characterization of small, mononuclear blood cells from Salmon having high phagocytic capacity and ability to differentiate into dendritic like cells. PLoS One. 2012;**7**(11):e49260

[25] Powsner RA, Edward ER. Essential Nuclear Physics. 2nd ed. India: Replika Press; 2006

[26] Ali U, Eze D, Ahmed L, Isa S, Ibrahim M. Endosulfan-induced changes in sperm count, testicular weight and some erythrocyte indices in male Guinea pigs. British Journal of Pharmacology and Toxicology. 2012;**3**(4):151-155

[27] Henry F. Fundamental of Radiation Protection. John Wiley & Sons Inc. New Jersey; 1969. pp. 111-117

[28] Franco HF, Helmut H, Bernard FG. Human basophil. Blood Journal. 2000;**96**(13):4028-4038

[29] El-shanshoury H, El-shanshoury G, Abaza A. Evaluation of low dose ionizing radiation effect on some blood components in animal model. Journal of Radiation Research and Applied Sciences. 2016;**9**:282-293

[30] Rozgaj R, Kasuba V, Sentija K, Prlic I. Radiation induced chromozomal aberation and haematological alteration in hospital. Occupational Medicine. 1999;**49**(6):353-360

[31] Ezz MK. The ameliorative effect of *Echinacea purpurea* against gamma radiation induced oxidative stress and immune responses in male rats. Australian Journal of Basic and Applied Sciences. 2011;**5**(10):506-512

[32] Seed TM, Fritz TE, Tolle DV, Jackson WE. Haematopoietic responses under protracted exposure to low daily dose gamma irradiation. Advances in Space Research. 2002;**30**(4): 945-955

[33] Robert F. Radiation Protection for Radiation Technologist. 2nd ed. New York: McGraw-Hill Book Company; 1996. pp. 18-30

Chapter 4

An Overview of PET Radiopharmaceuticals in Clinical Use: Regulatory, Quality and Pharmacopeia Monographs of the United States and Europe

Ya-Yao Huang

Additional information is available at the end of the chapter

http://dx.doi.org/10.5772/intechopen.79227

Abstract

Since 1976, more and more PET radiopharmaceuticals have been developed as the clinical introduction of [^{18}F]FDG for various medical applications. However, few of them could be involved in routinely clinical use in hospitals partly because of restrictions in regulatory and facilities. This chapter aims to provide an overview of PET radiopharmaceuticals that are common manufactured (or prepared) in industry (or hospitals) about regulatory and quality aspects, and further summarize pharmacopeia-listed PET radiopharmaceuticals and their clinical usefulness herein. Particularly, PET radiopharmaceuticals listed in latest United States Pharmacopeia (USP) and/or European Pharmacopeia (EP) are included for this chapter. Finally, this chapter would be helpful in the basic understanding of clinical PET radiopharmaceuticals for physicians or technologists.

Keywords: PET, radiopharmaceutical, regulation, quality, clinical application, USP, EP, pharmacopeia

1. Introduction

Positron emission tomography (PET) radiopharmaceutical is composed of a biologically active pharmacophore and a positron-emitting radionuclide, and belongs to a unique species in pharmaceutical field. The most common radionuclides for PET radiopharmaceuticals include ^{11}C, ^{15}O, ^{13}N, ^{18}F, ^{68}Ga and ^{82}Rb (**Table 1**). In addition to radiation issue, short half-lives of these positron emitters (78 sec~110 min) definitely result in unavoidable limitations on manufacturing (including production and following quality control (QC) analyses) and clinical use of PET radiopharmaceuticals. Above are all practical challenges for a conventional

Radionuclide	Half-life	Max specific activity (Ci/μmol)	ß⁺ (%)	Max $E_ß$ (MeV)	Max ß⁺ range (mm)	Production route
^{11}C	20 min	9220	99	0.96	4.1	Cyclotron
^{15}O	123 sec	90,800	100	1.19	5.1	Cyclotron
^{13}N	10 min	18,900	100	1.72	7.3	Cyclotron
^{18}F	110 min	1710	97	0.635	2.4	Cyclotron
^{68}Ga	68 min	2766	88	1.9	8.2	Cyclotron/ Generator
^{82}Rb	78 sec	150,400	95	3.35	14.1	Generator

Table 1. Characteristics of common positron emitters.

pharmaceutical industry. Hence, commercial large-scale manufacturing and small-scale preparation of PET radiopharmaceuticals are respectively allowed in radiopharmaceutical industries and the radiopharmacy of hospitals in most countries worldwide. Moreover, both practices in radiopharmaceutical industries and hospitals are clearly regulated by national competence authorities, such as Food and Drug Administration (FDA) of the United States (U.S.) and *European* Medicines Agency (EMA) of the European Union (EU).

In the other hand, a pharmacopeia is a national compendium of drug quality standards, such as U.S. Pharmacopeia (USP) and European Pharmacopeia (EP), and is always recognized as an official compendium. Drug standards listed in pharmacopeia monographs are usually enforced to be compliance under drug-related provisions at national level in order to prevent the marketing of inconsistent drugs and to reduce possible risks in public health. Although PET radiopharmaceuticals listing in pharmacopeia monographs sometimes do not mean for marketing authorization under national approval and reimbursement decision of medical insurance [1], some countries have enabled the clinical use (i.e., use for routine patient care with/without reimbursement or with/without national approval) or clinical trials as long as their qualities are in conformity with USP or EP standards, even no good manufacturing practice (*GMP*)-compliant process. Moreover, for those clinical studies using national-approved PET radiopharmaceutical for off-label indications, burdensome submission of an investigational new drug (IND) application will not be required in some countries.

In the other hand, specific QC procedures and specification of some PET radiopharmaceuticals have been listed in USP or EP. However, because of short half-lives of PET radiopharmaceuticals, QC tests prior to human administration within such a short period is a huge challenge. As a result, some quality exceptions are usually allowed for PET radiopharmaceuticals. Also, several efficient and quick tests have been developed for rapid QC tests of clinical PET radiopharmaceuticals.

This chapter first aims to provide an overview of regulations of manufacturing and clinical use of PET radiopharmaceuticals in U.S. and Europe. Secondly, the chapter will introduce the general quality aspect for PET radiopharmaceuticals. Finally, this chapter will end with the brief introduction of PET radiopharmaceuticals listed in the monographs of latest USP (USP 40) or EP (EP 9.0) (**Table 2**).

http://dx.doi.org/10.5772/intechopen.79227

Radionuclide	Compound	USP	EP
^{11}C	[^{11}C]CO	✓*	
	[^{11}C-methyl]Methionine	✓*	✓
	N-[^{11}C-methyl]Flumazenil	✓*	✓
	[^{11}C]N-methylspiroperidol	✓*	
	[^{11}C-methoxy]Raclopride	✓*	✓
	[1-^{11}C]Sodium Acetate	✓*	✓
^{13}N	[^{13}N]NH_3	✓	✓
^{15}O	[^{15}O]CO		✓
	[^{15}O]H_2O	✓*	✓
^{18}F	[^{18}F]FCH		✓
	[^{18}F]FDG	✓	✓
	[^{18}F]FDOPA (prepared by electrophilic substitution)	✓*	✓
	[^{18}F]FET		✓
	[^{18}F]FLT		✓
	[^{18}F]FMISO		✓
	[^{18}F]NaF	✓	✓
^{68}Ga	[^{68}Ga]Ga-Citrate		✓
	[^{68}Ga]Ga-DOTA-TOC		✓
^{82}Rb	[^{82}Rb]rubidium chloride	✓	

*These monographs of 8 FDA-unapproved PET radiopharmaceuticals have been omitted from USP since May 1, 2015 (USP 38).

Table 2. PET radiopharmaceuticals listed in USP and EP.

2. Regulatory aspects of PET radiopharmaceuticals in the USA and Europe

2.1. USA regulatory view

In U.S., the clinical use of all radiopharmaceuticals has been regulated by FDA since 1975. Briefly, the regulatory process can be divided into two types. They are: 1. IND submission for investigational and research purposes by an individual or a commercial manufacturer, and 2. submissions of Notice of Claimed Investigational Exemption (NCIE), an abbreviated new drug application (ANDA) or New Drug Application (NDA) for commercial marketing only by a commercial manufacturer. However, because of the increasing clinical need of PET radiopharmaceuticals, based on FDA Modernization Act (FDAMA) in 1997 [2], PET radiopharmaceuticals were first categorized as positron-emitting drugs. In the same time, all PET radiopharmaceutical manufacturing facilities in U.S. were programmatically to compliant with PET drug GMP-compliance guideline or with USP General Chapter <823> [3], and further registered as manufacturers. Till now, these legal manufacturers could on-site *(in-house)* produced PET radiopharmaceuticals with same specifications listed in USP monographs.

In the other hand, USP is annually published by a nonprofit organization since 1820, U.S. Pharmacopeial Convention, and such organization also worked with FDA and specialists in academia and companies to establish monographs or general chapters. Typically, USP monographs are typically developed after FDA approval of the drug product for commercial marketing and thus a USP monograph of an FDA-approved drug has been used as one basis for a reimbursement decision. The first USP monograph for a PET drug was published in 1990 [4] and it described the quality specification and analytic methods for [^{18}F]FDG injection. However, there had been an exception for 4 approved and 8 unapproved PET drugs listed in USP monographs till 2013. Moreover, not only these 12 monographs were provided to U.S. Pharmacopeial Convention by various academic sponsors with un-validated data and outdated analytic methods, but also these unapproved 8 PET drugs have limited commercial application without FDA-approved NDA or ANDA. Consequently, based on recommendations of the Society of Nuclear Medicine and Molecular Imaging (SNMMI) Committee [1], U.S. Pharmacopeial Convention announced the omission of the monographs of 8 unapproved PET drugs on June 2014 and the omission initiative became official on December 1, 2014.

2.2. European regulatory view

In Europe, radiopharmaceuticals have been recognized as a special group of medicines. Thus, the preparation and clinical use of PET radiopharmaceuticals have been regulated and variously adopted by member states. Similar to USP, EP has legal status in Europe. Compared to the USA, EP is only for drug quality and is independent of licensing status or clinical utility of such drug. Regarding to PET radiopharmaceuticals, corresponding monographs are elaborated by a group that is composed of academic, commercial and regulatory specialists. From another point of view, a number of EU member states have set up a regulatory framework from the definitions of "magistral and officinal formulae" that is listed in Article 3 of Directive 2001/83 [5]. Additionally, "in-house" small-scale preparation of PET radiopharmaceuticals is allowed without the requirements of a marketing authorization based on various national laws of European countries [5]. Both a general chapter of EP entitled "Extemporaneous Preparation of Radiopharmaceuticals" [6] and the new PIC/S guidance document with Annex 3 on radiopharmaceuticals [7] are published and worked as comprehensive guidelines for such magistral approach. Furthermore, because of the special characteristics of PET radiopharmaceuticals, the clinical studies using diagnostic radiopharmaceuticals do not fall within the GMP-compliance regulations of conventional drugs from EU Regulation no 536/2014 of 16 April 2014 [8, 9]. On brief summary, no matter EP or PIC/S document, they both clearly define a clear distinction between PET radiopharmaceuticals and conventional medicine, and further provide the corresponding guidance. All would be significantly helpful and powerful in promotion and development of PET radiopharmaceuticals in Europe.

3. Quality aspects of PET radiopharmaceuticals

Even costly implementation and maintenance of quality system for a PET radiopharmaceutical manufacturing (or preparing) site [10, 11], it is still thought to be cost-effective [12]. Moreover, it will be helpful for qualified patient care, regulatory requirements, optimization of safety and efficacy for patient care and a reliable quantitative performance in both diagnostic and therapeutic nuclear medicine procedures [13]. Therefore, GMP-compliant PET manufacturing (or preparing) process including production, QC, quality assurance (QA), package and distribution has been required by competent authorities in many countries worldwide. Furthermore, during these years, the concept of "Quality by Design (QbD)" based on guidelines of International Conference on Harmonization (ICH) (ICH Q8 [14], ICH Q9 [15], and ICH Q10 [16]) has been the fundamental topic in pharmaceutical field and an appropriate quality system has been widely required to implement in many radiopharmaceutical manufacturing sites (**Figure 1**). Briefly, QA covers whole process and GMP specifically characterizes those production and QC activities that guarantee products are produced under the constant scrutiny of quality standards [17], although the association of QA, GMP, and QC throughout whole pharmaceutical process is slightly different in various guidelines.

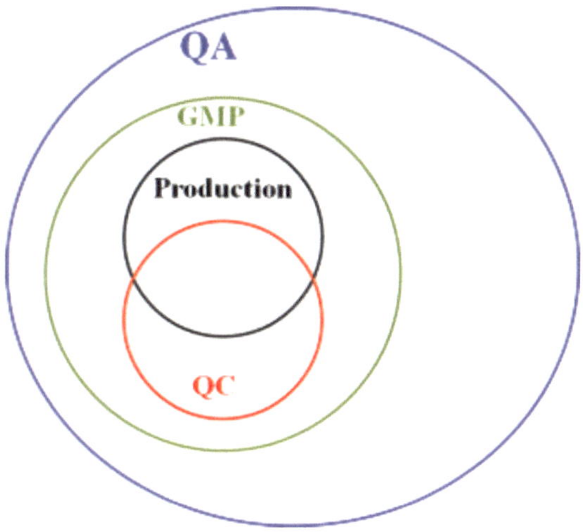

Figure 1. The inter-relationship for whole quality system in PET radiopharmaceutical manufacturing.

Particularly, QC procedure of PET radiopharmaceutical is usually critical and essential, since it is synthesized every day or is small-scale "prepared "in radiopharmacy of a hospital. A typical QC programme of a PET radiopharmaceutical is involved from radionuclide production to final product release and a series of QC tests for PET radiopharmaceuticals basically include:

1. Appearance, by visual assessment;
2. pH determination;
3. Radionuclidic identification, by gamma-ray spectrometry or half-life measurement;
4. Radionuclidic purity, by gamma-ray spectrometry;
5. Chemical purity, by high-pressure liquid chromatography (HPLC) or by thin-layer chromatography (TLC);
6. Radiochemical purity, by HPLC with a radioactivity detector or by TLC with a radioactivity scanner;

7. Residual solvents, by gas chromatography (GC);
8. Bacterial endotoxins, by a rabbit test or limulus amebocyte lysate (LAL) test;
9. Radioactivity, by a validated dose calibrator and.
10. Sterility, by incubating the sample with fluid thioglycollate medium (FTM) at 30~35°C for 14 days or with soybean casein digest (SCD) medium at 20~25°C for 14 days.

However, because of short-lives of PET radiopharmaceuticals, some lengthy tests cannot be performed prior to release for human use and are allowable to perform within a short time after the release. Furthermore, in addition to the limited time for QC of PET radiopharmaceuticals, limited personneal for *in-house* preparing of PET radiopharmaceuticals is another major issue for a hospital. Therefore, more and more efficient systems have been developed and successfully implemented for clinical use, such as Endosafe® Portable Testing System™ (PTS™) for rapid endotoxin testing (Charles River, Wilmington, MA) (https://www.criver.com/products-services/qc-microbial-solutions/endotoxin-testing/endotoxin-testing-systems/endosafe-nexgen-pts?region=3681) and Tracer-QC system for automation of QC tests of PET radiopharmaceuticals (LabLogic Systems Ltd., Sheffield, UK) (https://lablogic.com/nuclear-medicine-and-pet/instruments/tracer-qc).

4. Overview of current PET radiopharmaceuticals listed in USP or EP

4.1. [^{11}C-methyl]Methionine injection (EP)

Cellular protein synthesis is a well-control process for enzymes, membrane receptors, structural proteins, and growth factors [18]. Most importantly, increased cellular protein synthesis is often characterized in malignant growth [19]. Otherwise, decreased protein synthesis is found in certain neurodegenerative disorders [20]. Thus, the ability to *in vivo* visualize the protein synthesis rate is critical for clinic. Protein synthesis is initiated universally with the amino acid, methionine [21]. Therefore, one of ^{11}C-labeled methionine analogs, [^{11}C-methyl]methionine ([^{11}C]MET) [22] (**Figure 2**), has been used for imaging of rate of protein synthesis [23, 24], although the short physical half-life of ^{11}C (20 min) limits its accessibility for PET scanning centers without a cyclotron. Clinically, [^{11}C]MET has been used in imaging of brain, urinary, gynecological, liver and lung cancer [25–28]. Particularly, the enhanced transport of [^{11}C]MET into the brain has been known via the reversible sodium-independent transport system L (LAT 1) since 1995 [28] and increased LAT1 expression has been found in glioma and many other cancers and is associated with high grade and poor prognosis [29–32], thus [^{11}C]MET has been widely in various brain tumors [33, 34].

4.2. N-[^{11}C-methyl]Flumazenil injection (EP)

The GABA$_A$/benzodiazepine receptor complex is also known as the central benzodiazepine receptor and specifically mediates all pharmacologic properties of ethanol, zinc, picrotoxin and some drugs such as benzodiazepines (sedative, anxiolytic, anticonvulsant, myorelaxant),

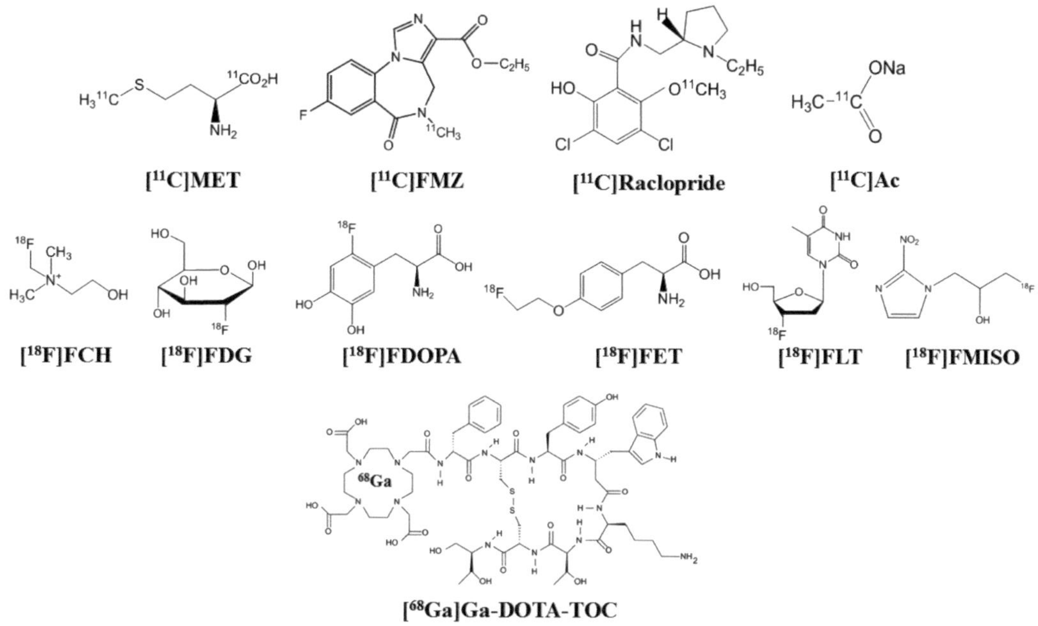

Figure 2. Chemical structures of PET radiopharmaceuticals listed in this chapter.

barbiturates (cerebral protection) and neuroactive steroids [35]. Based on a benzodiazepine antagonist, N-[^{11}C-methyl]Flumazenil ([^{11}C]FMZ) (**Figure 2**) [36] has been developed and known for its excellent kinetic properties for the image quantification [37]. Moreover, [^{11}C] FMZ has been considered as a versatile PET tracer for assessment of several conditions, such as neuronal damage in head injury [38], epilepsy [39], stroke-induced penumbral areas of infarction [40] and Alzheimer's disease (AD) [41].

4.3. [^{11}C-methoxy]Raclopride injection (EP)

Dopamine (DA) plays an important role in every-day brain functions including experiencing pleasure, regulating attention, and learning to control urges. Dysfunction of DA circuits has been thought to be related to various psychiatric diseases such as Parkinson's diseases (PD), addiction, attention-deficit hyperactivity disorder, and schizophrenia [42]. Studying *in vivo* dopamine function in humans became possible in the mid-1990s with the development of [^{11}C]raclopride (**Figure 2**) [43, 44], which originates from a DA receptor antagonist (D_2/D_3) with moderate affinity and reversible binding characteristics. Up to now, [^{11}C]raclopride is the most widely used PET radiopharmaceutical for measuring DA changes in striatal dopamine levels in the synapse before and after pharmacological and behavioral challenges [45], such as aging [46–48], schizophrenia [49–53] and PD [54, 55].

4.4. [1-^{11}C]sodium acetate injection (EP)

Acetate is a molecule quickly picked-up by cells to convert into acetyl-CoA by acetyl-CoA synthetase (EC 6.2.1.1 according to Enzyme Commission Number) and participates in cytoplasmic lipid synthesis, which is believed to be increased in tumors. Thus, [1-^{11}C] Sodium Acetate ([^{11}C]Ac) (**Figure 2**) [56, 57] has been proved clinical usefulness in prostate cancer (PC) [58], hepatocellular carcinoma (HCC), lung cancer, nasopharyngeal carcinoma [33], renal cell carcinoma, bladder carcinoma and brain tumors [59]. Furthermore, [^{11}C]Ac has been used to clinically measure myocardial oxygen consumption since 2010 [60] and used in some rare conditions, such as thymoma, cerebellopontine angle schwannoma, angiomyolipoma of the kidney, encephalitis, and multiple myeloma [59].

4.5. [^{13}N]NH_3 injection (USP and EP)

Coronary flow reserve (CFR) is calculated as the ratio of hyperemic to rest absolute myocardial blood flow (MBF) and is a particularly useful parameter in the assessment of adverse cardiovascular events such as epicardial coronary stenosis, diffuse atherosclerosis, and microvascular dysfunction on myocardial tissue perfusion [61]. Routinely used [^{13}N]Ammonia ([^{13}N]NH_3) is not only a useful ^{13}N-labeled PET imaging agent for assessing regional blood flow in tissues [62], but a well-validated radiotracer for clinical management of patients with coronary artery disease [62–64]. Moreover, recently [^{13}N]NH_3 has been used in PC, because the up-regulation of NH_3 during *de novo* glutamine synthesis was known in tumors [65]. Furthermore, because excess circulating NH_3 is neurotoxic and hyperammonemia is thought to be a major factor in the encephalopathy associated with several diseases, such as liver cirrhosis [66–68], [^{13}N]NH_3 is also used for elucidation of NH_3 metabolism in patients with hepatic encephalopathy [69].

4.6. [^{15}O]CO injection (EP)

[^{15}O]CO is one of the most common tracers used for noninvasively measuring oxygen consumption and blood volume [70, 71]. Additionally, [^{15}O]CO is crucial for the evaluation of acute stroke patients. Moreover, measurement of myocardial oxygen consumption is a useful tool to clarify the relationship between MBF and oxygen extraction fraction (OEF), because both OEF and MBF are important indicators in describing myocardial function [72].

4.7. [^{15}O]H_2O injection (EP)

Although the short half-life (123 sec) of ^{15}O results in the challenges in clinical use, [^{15}O]H_2O is still the preferred tracer because of its ease production from generator, effectiveness and safety for patient use [73]. Particularly, PET with [^{15}O]H_2O has been a standard method and most reliable approach for quantitative measurement of cerebral blood *flow* (*CBF*). Also, [^{15}O]H_2O is capable to clinically investigate cerebral and myocardial perfusion [74], and tumor perfusion [75, 76].

4.8. [^{18}F]FCH injection (EP)

Choline is a precursor for the biosynthesis of phospholipids which are essential components of all membranes and is phosphorylated by choline kinase (CK) to produce phosphatidylcholine. Upregulated CK is known in cancer cells, thus it further leads to increased uptake of choline in tumor cells with the excess need for phospholipid biosynthesis [77, 78]. Consequently, ^{18}F-labeled choline analogs, [^{18}F]fluoromethylcholine ([^{18}F]FCH) (**Figure 2**) [79, 80] has been a promising tumor imaging agents for various types of tumors include brain [80], breast, thyroid, lung, liver and prostate [81]. Particularly, [^{18}F]FCH has been shown to be better than [^{18}F]FDG for PC and HCC detections [81].

4.9. [^{18}F]FDG injection (USP and EP)

Since its synthesis in 1976, 2-fluorine-[^{18}F]fluorodeoxyglucose ([^{18}F]FDG) [82] (**Figure 2**) has been the most widely used radiotracer for PET studies in neuroscience, cardiology and oncology (**Table 3**) [83]. After FDA approval in 1997, [^{18}F]FDG with PET or PET/CT scanner became an established imaging tool in the clinical assessment of many neoplasms, as well as the nonmalignant diseases including dementia, myocardial ischaemia, inflammation and infection [84].

4.10. [^{18}F]FDOPA (prepared by electrophilic substitution) injection (EP)

Dihydroxyphenylalanine (DOPA) has been known as an intermediate in the catecholamine synthesis pathway. One of the ^{18}F-radiolabeled analogs, 3,4-dihydroxy-6-[^{18}F]fluoro-*L*-phenylalanine ([^{18}F]FDOPA) (**Figure 2**), was first reported as a PET tracer for imaging pre-synaptic dopaminergic functions in 1983 [85]. Subsequent studies revealed the utility of [^{18}F]FDOPA for the visualization of various peripheral tumor entities via PET [86], which can be attributed to the up-regulation of amino acid transporters in malignant tissues due to an often increased proliferation [87]. In particular, because of the relationship between the expression of aromatic L-amino acid decarboxylase (AADC) and the metabolism of [^{18}F]FDOPA [88, 89], [^{18}F]FDOPA has shown diagnostic advantages in the imaging of neuroendocrine cell-related malignancies like neuroendocrine tumors (NETs) [89–94], pheochromocytoma [95–97], pancreatic adenocarcinoma [98, 99] and neuroblastoma (NB) [100–102] regarding diagnostic efficiency and sensitivity.

Classification	Disease	Application
Neurology	Alzheimer's Disease	—
	Epilepsy	Pre-surgical evaluation for epileptogenic foci (85–90% accuracy).
Cardiology	Myocardial Viability	Assessment of myocardial viability prior to cardiac surgery
	Identify high-risk patients	Select patients who will benefit from bypass
Psychiatry	Schizophrenia	—
	Depression	—
Oncology	Tumor Evaluation	Differentiate recurrent/residual tumor from necrosis.
	Tumor Staging	Malignant vs. benign. Lung nodules, primary breast and colon cancers.
	Tumor Monitoring	Response to therapy.
	Tumor Localization	Metastases, abnormal sites
Infection and Inflammation	Orthopedic infections	—

Table 3. Summary for clinical application of [^{18}F]FDG [83].

4.11. [^{18}F]FET injection (EP)

Na^+-independent system L amino acid transporters (LATs) preferentially transports amino acids with large neutral side chains, including L-leucine, L-phenylalanine, and L-tyrosine. O-(2-[^{18}F] fluoroethyl)-L-tyrosine ([^{18}F]FET) (**Figure 2**) [103] belongs to the class of large neutral amino acids, which are transported via specific amino acid transporters especially of LATs [104]. Although data today still not reveal which the transporter(s) responsible for [^{18}F]FET accumulation in cells [105], [^{18}F]FET has been well known for its high uptake in brain tumors and its potential for grading tumors particularly gliomas [106, 107]. Summarily, [^{18}F]FET has been well-investigated in differential diagnosis, grading, prognostication, treatment response assessment, and differentiating pseudoprogression from non-specific post-therapeutic changes [108–110]. Switzerland was the first country to approve [^{18}F]FET PET for clinical use in brain tumor imaging since 2014 [105].

4.12. [^{18}F]FLT ([^{18}F]Alovudine) injection (EP)

Cellular proliferation plays an important role in cancer and has been an important imaging target of PET radiopharmaceuticals, especially with the aim targeting of DNA synthesis. Since the approach to the measurement of DNA synthesis in humans was explored in the early 1970s, based on an antiviral agent developed by Medivir, [^{18}F]fluorothymidine ([^{18}F]FLT, also known as [^{18}F]Alovudine) (**Figure 2**) [111, 112] has been designed with intracellularly trapping of its phosphorylated metabolite within cells [113]. Up to now, [^{18}F]FLT has been widely investigated in oncologic setting comprising tumor detection, staging, restaging, and response assessment to treatment [114–116] and [^{18}F]FLT imaging has several clinical advantages including noninvasive procedure, three-dimensional tumor images and simultaneous detection of multiple tumor sites [117]. Also, [^{18}F]FLT is capable to evaluate tumor heterogeneity in day-to-day practice [118].

4.13. [^{18}F]FMISO injection (EP)

Hypoxia means insufficient oxygen availability of a cell occurring both in health and is acknowledged by the observation of Gray *et al.* in the mid-1950s [119, 120]. Hypoxia is an important prognostic indicator of response to either chemotherapy or radiation therapy in cancer management [121, 122]. Hypoxia is also an independent factor for predicting the metastases tendency of a tumor cell, because of its enhancement in DNA mutations of atypical cells and further appearance of more aggressive cells. Consequently, 1-(2-hydroxy-3-[^{18}F]fluoropropyl)-2-nitroimidazole ([^{18}F]FMISO) (**Figure 2**) [123, 124] is the most established agent for assessing hypoxia and has been used for cancer imaging over the past 30 y for glioblastoma multiforme, non-small-cell lung cancer, and head and neck tumors [125]. In addition, high accuracy of [^{18}F]FMISO PET imaging for determining the duration of survival without relapses and for predicting the radiotherapy efficiency in patients with malignant tumors of various localizations has been reported [126, 127]. Furthermore, prognostic potential of [^{18}F]FMISO for the pretherapeutic tumor oxygenation status has been confirmed for glioblastoma multiforme, head and neck cancer, lung cancer, breast cancer, pancreatic cancer, gynecologic cancers, cervical cancer and sarcoma [127].

4.14. [^{18}F]NaF injection (USP and EP)

The bone is the most common place of tumor metastases next to the lung and liver [128]. Therefore, early and accurate diagnosis of the metastatic bone diseases thus plays an important role for an establishment of adequate therapeutic strategy [129]. [^{18}F]Sodium fluoride ([^{18}F]NaF) was introduced in 1962 and approved by FDA in 1972 [130]. [^{18}F]NaF is a high sensitive bone-seeking PET radiopharmaceutical and is considered as an excellent substitute for traditionally used ^{99m}Tc-labeled tracers, because its favorable characteristics of negligible protein binding, and rapid blood pool clearance. With ^{99m}Tc supply around the world is gradually become a crisis due to the shortage of ^{99}Mo-source material [131, 132], the clinical use of [^{18}F]NaF keeps increasing worldwide. Additionally, uptake of [^{18}F]NaF reflects blood flow and bone remodeling [133], and [^{18}F]NaF have been proposed for the use in detection of benign and malignant osseous abnormalities that also allows the regional characterization of lesions in metabolic bone diseases [134, 135].

4.15. [^{68}Ga]Ga-citrate injection(EP)

In addition to war and famine, bacterial infection has still been one of major worldwide causes for human morbidity and mortality for centuries [136, 137]. Because of the trapping of gallium in the extravascular compartment for inflammatory or infectious sites with the increased capillary permeability [138], and the iron-like binding characteristics in bacterial siderophores and activated lactoferrin in neutrophils [139, 140], gallium is thought to be indirectly uptaken by macrophages [141, 142] or directly uptaken by bacteria [143]. Thus, [^{67}Ga]gallium citrate ([^{68}Ga]Ga-Citrate) has been used for clinical imaging of infection and inflammation since 1984 [144]. The utilities of [^{68}Ga]Ga-Citrate include the monitoring of osteomyelitis, diskitis, intra-abdominal infection, tuberculosis and interstitial nephritis, as well as the localization of infection in patients with cellulitis and abscesses [145, 146].

4.16. [^{68}Ga]Ga-DOTA-TOC injection (EP)

NETs arised from neuroendocrine cells and are one of slow-growing tumors with year-by-year increased incidence rate and 75% of overall 5-y survival, which is strongly dependent on stage and grade of the tumor [147]. Because NETs has been known for its unique over-expression of somatostatin receptors (SSTrs) on the tumor cells [148], SSTr-targeting PET radiopharmaceuticals provide a promising and useful approach for both diagnostic imaging and further peptide receptor radionuclide therapy (PRRT), such as ^{68}Ga-labeled DOTA-(Tyr^{3})-octreotide acetate ([^{68}Ga]Ga-DOTA-TOC) (**Figure 2**) [149]. Because octreotide is a subset of the amino acid in somatostatin and has been demonstrated to avidly bind to SSTr [150], [^{68}Ga]Ga-DOTA-TOC has been recognized for its affinity toward both the type 2 somatostatin receptor (SSTr2) and the type 5 somatostatin receptor (SSTr5) [151–154]. Also, [^{68}Ga]Ga-DOTA-TOC was the first PET radiopharmaceutical to clinically localize to NETs in 2001 [155] and has been widely used in Europe and several other countries to assist the therapy planning and accurate diagnosis of NETs patients [156]. In addition, [^{68}Ga]Ga-DOTA-TOC is valuable for neuroectodermal tumors, Hurthle cell thyroid carcinoma, prostate cancer patients with bone metastases and autoimmune thyroid disease like Graves' disease and Hashimoto's disease [145, 146].

4.17. [^{82}Rb]rubidium chloride (USP)

Just like previous described [^{13}N]NH_3 and [^{15}O]H_2O, [^{82}Rb]Rubidium chloride ([^{82}Rb]RbCl) has been reported for directly proportional relationship between its uptake and MBF since 1954 [157]. In addition, several studies have demonstrated the good diagnostic accuracy of [^{82}Rb]RbCl in monitoring of cardiac flow [158, 159]. Subsequently, $^{82}Sr/^{82}Rb$ generator (CardioGen-82®) of Bracco Diagnostics has been approved by FDA for clinical cardiac imaging since 1989 (NDA 19–414). Therefore, production and administration of [^{82}Rb]RbCl can be well coordinated with the $^{82}Sr/^{82}Rb$ generator in clinic [160], although a short half-life (78 sec) of ^{82}Rb. In brief, the clinical advantages of [^{82}Rb]RbCl cardiac imaging include its capacity to accurately quantify MBF and a low delivered radiation exposure for a rest/stress test resulted from its very short half-life [160].

5. Conclusion

With the development of imaging technology, more and more pharmaceutical industry and hospitals worldwide have paid attentions on clinical potential of PET radiopharmaceuticals. However, because of special characteristics of PET radiopharmaceuticals, current pharmaceutical regulatory is probably inapplicable and would be a hurdle for clinical use of PET radiopharmaceuticals in most countries. Thus, as these official monographs of PET radiopharmaceuticals listing in USP or EP, it is definitely worthy to work together for more pharmacopeia monographs and a PET radiopharmaceutical-specific regulatory for benefits of patient-centered care in the future.

Acknowledgements

This work has been supported in part by grants from the National Taiwan University Hospital, Grants NTUH107-S3882.

Conflict of interest

We declare no conflict of interest.

Author details

Ya-Yao Huang

Address all correspondence to: careyyh@ntuh.gov.tw

PET Center, Department of Nuclear Medicine, National Taiwan University Hospital, Taipei, Taiwan

References

[1] Schwarz S, Norenberg J, Berridge M, et al. The future of USP monographs for PET drugs. Journal of Nuclear Medicine. 2013;**54**(3):472-475

[2] Food and Drug Administration Modernization Act of 1997. Washington, DC: Congress US; 1997

[3] USP. Positron emission tomography drugs for compounding, investigational, and research uses, USP<823>. In: The United States Pharmacopeia, 40th Rev., and the National Formulary, 35th Ed. 40 Ed. Rockville, MD: United States Pharmacopeial Convention, Inc.; 2017. pp. 758-768

[4] U.S. Pharmacopeial Convention. USP 22–NF 17, Fludeoxyglucose F 18 Injection. Rockville, MD: U.S. Pharmacopeial Convention; 1990:579-580

[5] Decristoforo C, Patt M. Are we "preparing" radiopharmaceuticals? EJNMMI Radiopharmacy and Chemistry. 2017;**2**(1):12

[6] EP9.0. 5.19. Extemporaneous preparation of radiopharmaceuticals. European Pharmacopeia. 8th ed. Strasbourg, France United States Pharmacopeial Convention, Inc.; 2017:767-772

[7] Pharmaceutical Inspection Co-operation Scheme. PIC/S guide to good practices for the preparation of medicinal products in healthcare establishments. PE 010-4. Geneva: Pharmaceutical Inspection Co-operation Scheme; 2014. Available from: www.picscheme.org

[8] Decristoforo C, Penuelas I, Elsinga P, et al. Radiopharmaceuticals are special, but is this recognized? The possible impact of the new clinical trials regulation on the preparation of radiopharmaceuticals. European Journal of Nuclear Medicine and Molecular Imaging. 2014;**41**(11):2005-2007

[9] Ballinger JR, Koziorowski J. Regulation of PET radiopharmaceuticals production in Europe. In: Khalil MM, editor. Basic Science of PET Imaging. Egypt: Springer; 2017. pp. 127-143

[10] Bergmann H, Busemann-Sokole E, Horton P. Quality assurance and harmonisation of nuclear medicine investigations in Europe. European Journal of Nuclear Medicine. 1995;**22**(5):477-480

[11] Liu K-T, Zhao J-H, Men L-C, Chen C-H. Quality by design and risk assessment for radiopharmaceutical manufacturing and clinical imaging. In: Akyar I, editor. Latest Research into Quality Control. IntechOpen; 2012. pp. 255-292. DOI: 10.5772/51112. Available from: https://www.intechopen.com/books/latest-research-into-quality-control/quality-by-design-and-risk-assessment-for-radiopharmaceutical-manufacturing-and-clinical-imaging

[12] Chuck A, Jacobs P, Logus J, St Hilaire D, Chmielowiec C, McEwan A. Marginal cost of operating a positron emission tomography center in a regulatory environment. International Journal of Technology Assessment in Health Care. 2005;**21**(4):442-451

[13] Zimmerman B, Herbst C, Norenberg J, Woods M. International guidance on the establishment of quality assurance programmes for radioactivity measurement in nuclear medicine. Applied Radiation and Isotopes. 2006;**64**(10-11):1142-1146

[14] International Conference on Harmonisation. ICH Harmonised Tripartite Guideline: Pharmaceutical Development Q8 (R2). http://www.ich.org/fileadmin/Public_Web_Site/ICH_Products/Guidelines/Quality/Q8_R1/Step4/Q8_R2_Guideline.pdf

[15] International Conference on Harmonisation. ICH Harmonised Tripartite Guideline, Quality Risk Management, Q9. http://www.ich.org/fileadmin/Public_Web_Site/ICH_Products/Guidelines/Quality/Q9/Step4/Q9_Guideline.pdf

[16] International Conference on Harmonisation. ICH Harmonised Tripartite Guideline, Pharmaceutical Quality System, Q10. http://www.ich.org/fileadmin/Public_Web_Site/ICH_Products/Guidelines/Quality/Q10/Step4/Q10_Guideline.pdf

[17] US Pharmacopeial Convention. Ensuring the Quality of Medicines in Resource-Limited Countries. https://www.usp-pqm.org/sites/default/files/pqms/article/ensuringquality-operationalguide.pdf

[18] Sonenberg N, Hinnebusch AG. Regulation of translation initiation in eukaryotes: Mechanisms and biological targets. Cell. 2009;**136**(4):731-745

[19] Bhat M, Robichaud N, Hulea L, Sonenberg N, Pelletier J, Topisirovic I. Targeting the translation machinery in cancer. Nature Reviews. Drug Discovery. 2015;**14**(4):261-278

[20] Halliday M, Mallucci GR. Modulating the unfolded protein response to prevent neurodegeneration and enhance memory. Neuropathology and Applied Neurobiology. 2015;**41**(4):414-427

[21] Kozak M. Comparison of initiation of protein synthesis in procaryotes, eucaryotes, and organelles. Microbiological Reviews. 1983;**47**(1):1-45

[22] Langstrom B, Antoni G, Gullberg P, et al. Synthesis of L-and D-[methyl-^{11}C] methionine. Journal of Nuclear Medicine. 1987;**28**(6):1037-1040

[23] Hawkins R, Huang S, Barrio J, et al. Estimation of local cerebral protein synthesis rates with L-[1-^{11}C] leucine and PET: Methods, model, and results in animals and humans. Journal of Cerebral Blood Flow and Metabolism. 1989;**9**(4):446-460

[24] Ishiwata K, Vaalburg W, Elsinga P, Paans A, Woldring M. Comparison of L-[1-^{11}C] methionine and L-methyl-[^{11}C] methionine for measuring in vivo protein synthesis rates with PET. Journal of Nuclear Medicine. 1988;**29**(8):1419-1427

[25] Nunez R, Macapinlac H, Yeung H, et al. Combined ^{18}F-FDG and ^{11}C-methionine PET scans in patients with newly progressive metastatic prostate cancer. Journal of Nuclear Medicine. 2002;**43**(1):46-55

[26] Wood K, Hoskin P, Saunders M. Positron emission tomography in oncology: A review. Clinical Oncology (Royal College of Radiologists). 2007;**19**(4):237-255

[27] Kuang Y, Wang F, Corn DJ, Tian H, Lee Z. Metabolism of Radiolabeled Methionine in Hepatocellular carcinoma. Molecular Imaging and Biology. 2014;**16**(1):44-52

[28] Lilja A, Bergström K, Hartvig P, et al. Dynamic study of supratentorial gliomas with L-methyl-^{11}C-methionine and positron emission tomography. AJNR. American Journal of Neuroradiology. 1985;**6**(4):505-514

[29] Nawashiro H, Otani N, Shinomiya N, et al. L-type amino acid transporter 1 as a potential molecular target in human astrocytic tumors. International Journal of Cancer. 2006;**119**(3):484-492

[30] Betsunoh H, Fukuda T, Anzai N, et al. Increased expression of system large amino acid transporter (LAT)-1 mRNA is associated with invasive potential and unfavorable prognosis of human clear cell renal cell carcinoma. BMC Cancer. 2013;**13**:509

[31] Kaira K, Sunose Y, Arakawa K, et al. Prognostic significance of L-type amino-acid transporter 1 expression in surgically resected pancreatic cancer. British Journal of Cancer. 2012;**107**(4):632-638

[32] Takeuchi K, Ogata S, Nakanishi K, et al. LAT1 expression in non-small-cell lung carcinomas: Analyses by semiquantitative reverse transcription-PCR (237 cases) and immunohistochemistry (295 cases). Lung Cancer. 2010;**68**(1):58-65

[33] Pantaleo M, Nannini M, Maleddu A, et al. Conventional and novel PET tracers for imaging in oncology in the era of molecular therapy. Cancer Treatment Reviews. 2008;**34**(2):103-121

[34] Juhász C, Dwivedi S, Kamson DO, Michelhaugh SK, Mittal S. Comparison of amino acid positron emission Tomographic radiotracers for molecular imaging of primary and metastatic brain tumors. Molecular Imaging. 2014;**13**. DOI: 10.2310/7290.2014.00015

[35] Heiss W, Herholz K. Brain receptor imaging. Journal of Nuclear Medicine. 2006; **47**(2):302-312

[36] Maziere M, Prenant C, Sastre J, et al. ^{11}C-RO-15-1788 et ^{11}C-flunitrazepam, deux coordinats pour l'etude par tomographie par positrons des sites de liaison des benzodiazepines. Comptes rendus de l'Académie des Sciences (Paris). 1983;**296**:871-876

[37] Koeppe R, Holthoff V, Frey K, Kilbourn M, Kuhl D. Compartmental analysis of [^{11}C] flumazenil kinetics for the estimation of ligand transport rate and receptor distribution using positron emission tomography. Journal of Cerebral Blood Flow and Metabolism. 1991;**11**(5):735-744

[38] Geeraerts T, Coles J, Aigbirhio F, et al. Validation of reference tissue modelling for [^{11}C] flumazenil positron emission tomography following head injury. Annals of Nuclear Medicine. 2011;**25**(6):396-405

[39] Lamusuo S, Pitkänen A, Jutila L, et al. [^{11}C]Flumazenil binding in the medial temporal lobe in patients with temporal lobe epilepsy: Correlation with hippocampal MR volumetry, T2 relaxometry, and neuropathology. Neurology. 2000;**54**(12):2252-2260

[40] Heiss W, Sobesky J, Smekal U, et al. Probability of cortical infarction predicted by flumazenil binding and diffusion-weighted imaging signal intensity: A comparative positron emission tomography/magnetic resonance imaging study in early ischemic stroke. Stroke. 2004;**35**(8):1892-1898

[41] Pascual B, Prieto E, Arbizu J, et al. Decreased carbon-11-flumazenil binding in early Alzheimer's disease. Brain. 2012;**135**(Pt 9):2817-2825

[42] Dichter GS, Damiano CA, Allen JA. Reward circuitry dysfunction in psychiatric and neurodevelopmental disorders and genetic syndromes: Animal models and clinical findings. Journal of Neurodevelopmental Disorders. 2012;**4**(1):19

[43] Ehrin E, Gawell L, Hogberg T, de Paulis T, Strom P. Synthesis of [methoxy-^{3}H]-and [methoxy-^{11}C]-labelled raclopride. Specific dopamine-D_2 receptor ligands. Journal of Labelled Compounds and Radiopharmaceuticals. 1987;**24**(8):931-940

[44] Farde L, Ehrin E, Eriksson L, et al. Substituted benzamides as ligands for visualization of dopamine receptor binding in the human brain by positron emission tomography. Proceedings of the National Academy of Sciences of the United States of America. 1985;**82**(11):3863-3867

[45] Egerton A, Mehta MA, Montgomery AJ, et al. The dopaminergic basis of human behaviors: A review of molecular imaging studies. Neuroscience and Biobehavioral Reviews. 2009;**33**(7):1109-1132

[46] Pohjalainen T, Rinne J, Nagren K, SyvAlahti E, Hietala J. Sex differences in the striatal dopamine D_2 receptor binding characteristics in vivo. American Journal of Psychiatry Research. 1998;**155**(6):768-773

[47] Volkow N, Wang G, Fowler J, et al. Measuring age-related changes in dopamine D_2 receptors with ^{11}C-raclopride and ^{18}F-N-methylspiroperidol. Psychiatry Research. 1996; **67**(1):11-16

[48] Vernaleken I, Weibrich C, Siessmeier T, et al. Asymmetry in dopamine $D_{2/3}$ receptors of caudate nucleus is lost with age. NeuroImage. 2007;**34**(3):870-878

[49] Wong D, Wagner H Jr, Tune L, et al. Positron emission tomography reveals elevated D_2 dopamine receptors in drug-naive schizophrenics. Science. 1986;**234**(4783):1558-1563

[50] Farde L, Wiesel F, Stone-Elander S, et al. D_2 dopamine receptors in neuroleptic-naive schizophrenic patients. A positron emission tomography study with [^{11}C]raclopride. Archives of General Psychiatry. 1990;**47**:213-219

[51] Buchsbaum M, Christian B, Lehrer D, et al. D_2/D_3 dopamine receptor binding with [^{18}F]fallypride in thalamus and cortex of patients with schizophrenia. Schizophrenia Research. 2006;**85**(1-3):232-244

[52] Talvik M, Nordstrom A, Olsson H, Halldin C, Farde L. Decreased thalamic D_2/D_3 receptor binding in drug-naive patients with schizophrenia: A PET study with [^{11}C]FLB457. The International Journal of Neuropsychopharmacology. 2003;**6**(04):361-370

[53] Yasuno F, Suhara T, Okubo Y, et al. Low dopamine D_2 receptor binding in subregions of the thalamus in schizophrenia. The American Journal of Psychiatry. 2004;**161**(6):1016-1022

[54] Kaasinen V, Nagren K, Hietala J, et al. Extrastriatal dopamine D_2 and D_3 receptors in early and advanced Parkinson's disease. Neurology. 2000;**54**(7):1482-1487

[55] Kaasinen V, Aalto S, Nagren K, Hietala J, Sonninen P, Rinne J. Extrastriatal dopamine D_2 receptors in Parkinson's disease: A longitudinal study. Journal of Neural Transmission. 2003;**110**(6):591-601

[56] Shreve P, Chiao P, Humes H, Schwaiger M, Gross M. Carbon-11-acetate PET imaging in renal disease. Journal of Nuclear Medicine. 1995;**36**(9):1595-1601

[57] Yeh S, Liu R, Wu L, Yen S, Chang C, Chen K. ^{11}C-acetate clearance in nasopharyngeal carcinoma. Nuclear Medicine Communications. 1999;**20**(2):131-134

[58] Jadvar H. Prostate Cancer: PET with ^{18}F-FDG, ^{18}F-or ^{11}C-acetate, and ^{18}F-or ^{11}C-Choline. Journal of Nuclear Medicine. 2011;**52**(1):81-89

[59] Grassi I, Nanni C, Allegri V, et al. The clinical use of PET with ^{11}C-acetate. American Journal of Nuclear Medicine and Molecular Imaging. 2012;**2**(1):33-47

[60] Sörensen J, Valind S, Andersson L. Simultaneous quantification of myocardial perfusion, oxidative metabolism, cardiac efficiency and pump function at rest and during supine bicycle exercise using 1-^{11}C-acetate PET--a pilot study. Clinical Physiology and Functional Imaging. 2010;**30**(4):279-284

[61] Gould KL, Johnson NP, Bateman TM, et al. Anatomic versus physiologic assessment of coronary artery disease: Role of coronary flow reserve, fractional flow reserve, and positron emission tomography imaging in revascularization decision-making. Journal of the American College of Cardiology. 2013;**62**(18):1639-1653

[62] Phelps M, Hoffman E, Raybaud C. Factors which affect cerebral uptake and retention of ^{13}NH3. Stroke. 1977;**8**(6):694-702

[63] Walsh W, Fill H, Harper P. Nitrogen-13-labeled ammonia for myocardial imaging. Seminars in Nuclear Medicine. 1977;**7**(1):59-66

[64] Harper P, Schwartz J, Beck R, et al. Clinical myocardial imaging with nitrogen-13 ammonia. Radiology. 1973;**108**(3):613-617

[65] Shi X, Zhang X, Yi C, Liu Y. He Q. [^{13}N] Ammonia positron emission tomographic/computed tomographic imaging targeting glutamine synthetase expression in prostate cancer. Molecular Imaging. 2014;**13**:1-10

[66] Häberle J. Clinical practice: The management of hyperammonemia. European Journal of Pediatrics. 2011;**170**(1):21-34

[67] Butterworth R. Pathophysiology of brain dysfunction in hyperammonemic syndromes: The many faces of glutamine. Molecular Genetics and Metabolism. 2014;**113**(1-2):113-117

[68] Hadjihambi A, Khetan V, Jalan R. Pharmacotherapy for hyperammonemia. Expert Opinion on Pharmacotherapy. 2014;**15**(12):1685-1695

[69] Keiding S, Pavese N. Brain metabolism in patients with hepatic encephalopathy studied by PET and MR. Archives of Biochemistry and Biophysics. 2013;**536**(2):131-142

[70] Kobayashi M, Kudo T, Tsujikawa T, et al. Shorter examination method for the diagnosis of misery perfusion with count-based oxygen extraction fraction elevation in ^{15}O-gas PET. Journal of Nuclear Medicine. 2008;**49**(2):242-246

[71] Shidahara M, Watabe H, Kim K, Kudomi N, Ito H, Iida H. Optimal scan time of oxygen-15-labeled gas inhalation autoradiographic method for measurement of cerebral oxygen extraction fraction and cerebral oxygen metabolic rate. Annals of Nuclear Medicine. 2008;**22**(8):667-675

[72] Iida H, Rhodes C, Araujo L, et al. Noninvasive quantification of regional myocardial metabolic rate for oxygen by use of $^{15}O_2$ inhalation and positron emission tomography. Theory, error analysis, and application in humans. Circulation. 1996;**94**(4):792-807

[73] Reutens D, Bittar R, Tochon-Danguy H, Scott A. Clinical applications of [^{15}O]H_2O PET activation studies. Clinical Positron Imaging. 1999;**2**(3):145-152

[74] Ter-Pogossian M, Herscovitch P. Radioactive oxygen-15 in the study of cerebral blood flow, blood volume, and oxygen metabolism. Seminars in Nuclear Medicine. 1985;**15**(4):377-394

[75] Josephs D, Spicer J, O'Doherty M. Molecular imaging in clinical trials. Targeted Oncology. 2009;**4**(3):151-168

[76] Boss D, Olmos R, Sinaasappel M, Beijnen J, Schellens J. Application of PET/CT in the development of novel anticancer drugs. The Oncologist. 2008;**13**(1):25-38

[77] de Molina AR, Rodríguez-González AN, Gutiérrez R, et al. Overexpression of choline kinase is a frequent feature in human tumor-derived cell lines and in lung, prostate, and colorectal human cancers. Biochemical and Biophysical Research Communications 2002;**296**(3):580-583

[78] Ramírez dMA, Penalva V, Lucas L, Lacal J. Regulation of choline kinase activity by Ras proteins involves Ral-GDS and PI3K. Oncogene. 2002;**21**(6):937-946

[79] DeGrado TR, Coleman RE, Wang S, et al. Synthesis and evaluation of ^{18}F-labeled choline as an oncologic tracer for positron emission tomography: Initial findings in prostate cancer. Cancer Research. 2000;**61**(1):110-117

[80] DeGrado TR, Baldwin SW, Wang S, et al. Synthesis and evaluation of ^{18}F-labeled choline analogs as oncologic PET tracers. Journal of Nuclear Medicine. 2001;**42**(12):1805-1814

[81] Huang Y, Tsai C, Wen H, Tzen K, Yen R, Shiue C. High yield one-pot production of [^{18}F]FCH via a modified TRACERlab Fx_{FN} module. Applied Radiation and Isotopes. 2017;**128**:190-198

[82] Ido T, Wan CN, Casella V, et al. Labeled 2-deoxy-D-glucose analogs. ^{18}F-labeled 2-deoxy-2-fluoro-*D*-glucose, 2-deoxy-2-fluoro-*D*-mannose and ^{14}C-2-deoxy-2-fluoro-*D*-glucose. Journal of Labelled Compounds and Radiopharmaceuticals. 1978;**14**(2):175-183

[83] Shiue CY. Development and design of radiopharmaceuticals: (II) Radiolabelling. In: Paper Presented at: New Radiotracer Development: From Bench to Bedside; 2009/08/15. Taichung, Taiwan: The Taiwanese Society of Medical Cyclotron; 2009

[84] Kitson SL, Cuccurullo V, Ciarmiello A, Salvo D, Mansi L. Clinical applications of positron emission tomography (PET) imaging in medicine: Oncology, brain diseases and cardiology. Current Radiopharmaceuticals. 2009;**2**(4):224-253

[85] Garnett E, Firnau G, Nahmias C. Dopamine visualized in the basal ganglia of living man. Nature. 1983;**305**(5930):137-138

[86] Seibyl J, Chen W, Silverman D. 3, 4-dihydroxy-6-[^{18}F]-fluoro-*L*-phenylalanine positron emission tomography in patients with central motor disorders and in evaluation of brain and other tumors. Seminars in Nuclear Medicine. 2007;**37**(6):440-450

[87] Isselbacher K. Sugar and amino acid transport by cells in culture--differences between normal and malignant cells. The New England Journal of Medicine. 1972;**286**(17):929-933

[88] Koopmans K, Neels O, Kema I, et al. Molecular imaging in neuroendocrine tumors: Molecular uptake mechanisms and clinical results. Critical Reviews in Oncology/Hematology. 2009;**71**(3):199-213

[89] Neels O, Koopmans K, Jager P, et al. Manipulation of [^{11}C]-5-hydroxytryptophan and 6-[^{18}F] fluoro-3, 4-dihydroxy-L-phenylalanine accumulation in neuroendocrine tumor cells. Cancer Research. 2008;**68**(17):7183-7190

[90] Minn H, Kauhanen S, Seppänen M, Nuutila P. ^{18}F-FDOPA: a multiple-target molecule. Journal of Nuclear Medicine. 2009;**50**(12):1915-1918

[91] Jager PL, Chirakal R, Marriott CJ, Brouwers AH, Koopmans KP, Gulenchyn KY. 6-*L*-^{18}F-fluorodihydroxyphenylalanine PET in neuroendocrine tumors: Basic aspects and emerging clinical applications. Journal of Nuclear Medicine. 2008;**49**(4):573-586

[92] Balogova S, Talbot J-N, Nataf V, et al. ^{18}F-Fluorodihydroxyphenylalanine vs other radiopharmaceuticals for imaging neuroendocrine tumours according to their type. European Journal of Nuclear Medicine and Molecular Imaging. 2013;**40**(6):943-966

[93] Chondrogiannis S, Grassetto G, Marzola M, et al. ^{18}F-DOPA PET/CT biodistribution consideration in 107 consecutive patients with neuroendocrine tumours. Nuclear Medicine Communications. 2012;**33**(2):179-184

[94] Rufini V, Treglia G, Montravers F, Giordano A. Diagnostic accuracy of [^{18}F]DOPA PET and PET/CT in patients with neuroendocrine tumors: A meta-analysis. Clinical and Translational Imaging. 2013;**1**(2):1-12

[95] Rischke H, Benz M, Wild D, et al. Correlation of the genotype of paragangliomas and pheochromocytomas with their metabolic phenotype on 3, 4-dihydroxy-6-^{18}F-fluoro-L-phenylalanin PET. Journal of Nuclear Medicine. 2012;**53**(9):1352-1358

[96] Gabriel S, Blanchet E, Sebag F, et al. Functional characterization of nonmetastatic paraganglioma and pheochromocytoma by 18F-FDOPA PET: Focus on missed lesions. Clinical Endocrinology. 2013;**79**(2):170-177

[97] Marzola M, Chondrogiannis S, Grassetto G, et al. 18F-DOPA PET/CT in the evaluation of hereditary SDH-deficiency Paraganglioma-Pheochromocytoma syndromes. Clinical Nuclear Medicine. 2014;**39**(1):e53-e58

[98] Tuomela J, Forsback S, Haavisto L, et al. Enzyme inhibition of dopamine metabolism alters 6-[18F]FDOPA uptake in orthotopic pancreatic adenocarcinoma. EJNMMI Research. 2013;**3**:18

[99] Jadvar H. Hepatocellular carcinoma and gastroenteropancreatic neuroendocrine tumors: Potential role of other positron emission tomography radiotracers. Seminars in Nuclear Medicine. 2012;**42**(4):247-254

[100] Piccardo A, Lopci E, Conte M, et al. Bone and lymph node metastases from Neuroblastoma detected by ^{18}F-DOPA-PET/CT and confirmed by Posttherapy ^{131}I-MIBG but negative on diagnostic ^{123}I-MIBG scan. Clinical Nuclear Medicine. 2014;**39**(1):e80-e83

[101] Lopci E, Piccardo A, Nanni C, et al. ^{18}F-DOPA PET/CT in neuroblastoma: Comparison of conventional imaging with CT/MR. Clinical Nuclear Medicine. 2012;**37**(4):e73-e78

[102] Lu MY, Liu YL, Chang HH, et al. Characterization of Neuroblastic tumors using ^{18}F-FDOPA PET. Journal of Nuclear Medicine. 2013;**54**:1-8

[103] Wester H, Herz M, Weber W, et al. Synthesis and radiopharmacology of O-(2-[^{18}F] fluoroethyl)-L-tyrosine for tumor imaging. Journal of Nuclear Medicine. 1999;**40**(1): 205-212

[104] McConathy J, Yu W, Jarkas N, Seo W, Schuster D, Goodman M. Radiohalogenated nonnatural amino acids as PET and SPECT tumor imaging agents. Medicinal Research Reviews. 2012;**32**(4):868-905

[105] Langen K, Stoffels G, Filss C, et al. Imaging of amino acid transport in brain tumours: Positron emission tomography with O-(2-[^{18}F] fluoroethyl)-L-tyrosine (FET). Methods. 2017;**130**:124-134

[106] Pöpperl G, Kreth F, Mehrkens J, et al. FET PET for the evaluation of untreated gliomas: Correlation of FET uptake and uptake kinetics with tumour grading. European Journal of Nuclear Medicine and Molecular Imaging. 2007;**34**(12):1933-1942

[107] Pauleit D, Floeth F, Tellmann L, et al. Comparison of O-(2-^{18}F-fluoroethyl)-L-tyrosine PET and 3-^{123}I-iodo-alpha-methyl-L-tyrosine SPECT in brain tumors. Journal of Nuclear Medicine. 2004;**45**(3):374-381

[108] Floeth F, Sabel M, Stoffels G, et al. Prognostic value of ^{18}F-fluoroethyl-L-tyrosine PET and MRI in small nonspecific incidental brain lesions. Journal of Nuclear Medicine. 2008;**49**(5):730-737

[109] Piroth M, Pinkawa M, Holy R, et al. Prognostic value of early [^{18}F]fluoroethyltyrosine positron emission tomography after radiochemotherapy in glioblastoma multiforme. International Journal of Radiation Oncology, Biology, Physics. 2011;**80**(1):176-184

[110] Weckesser M, Langen K, Rickert C, et al. O-(2-[^{18}F] fluorethyl)-L-tyrosine PET in the clinical evaluation of primary brain tumours. European Journal of Nuclear Medicine and Molecular Imaging. 2005;**32**(4):422-429

[111] Grierson J, Shields A. Radiosynthesis of 3′-deoxy-3′-[^{18}F] fluorothymidine:[^{18}F] FLT for imaging of cellular proliferation in vivo. Nuclear Medicine and Biology. 2000; **27**(2):143-156

[112] Shields A. PET imaging with ^{18}F-FLT and thymidine analogs: Promise and pitfalls. Journal of Nuclear Medicine. 2003;**44**(9):1432-1434

[113] Kong X, Zhu Q, Vidal P, et al. Comparisons of anti-human immunodeficiency virus activities, cellular transport, and plasma and intracellular pharmacokinetics of 3′-fluoro-3′-deoxythymidine and 3′-azido-3′-deoxythymidine. Antimicrobial Agents and Chemotherapy. 1992;**36**(4):808-818

[114] Tehrani OS, Shields AF. PET imaging of proliferation with pyrimidines. Journal of Nuclear Medicine. 2013;**54**(6):903-912

[115] Herrmann K, Buck AK. Proliferation imaging with ^{18}F-Fluorothymidine PET/computed tomography: Physiologic uptake, variants, and pitfalls. PET Clinics. 2014;**9**(3):331-338

[116] Bollineni V, Kramer G, Jansma E, Liu Y, Oyen W. A systematic review on [^{18}F] FLT-PET uptake as a measure of treatment response in cancer patients. European Journal of Cancer. 2016;**55**:81-97

[117] Everitt S, Ball D, Hicks R, et al. Differential ^{18}F-FDG and ^{18}F-FLT uptake on serial PET/CT imaging before and during definitive Chemoradiation for non-small cell lung Cancer. Journal of Nuclear Medicine. 2014;**55**(7):1069-1074

[118] Herrmann K, Wieder HA, Buck AK, et al. Early response assessment using 3′-deoxy-3′-[^{18}F] fluorothymidine-positron emission tomography in high-grade non-Hodgkin's lymphoma. Clinical Cancer Research. 2007;**13**(12):3552-3558

[119] Thomlinson R, Gray L. The histological structure of some human lung cancers and the possible implications for radiotherapy. British Journal of Cancer. 1955;**9**(4):539-549

[120] Gray L, Conger A, Ebert M, Hornsey S, Scott O. The concentration of oxygen dissolved in tissues at the time of irradiation as a factor in radiotherapy. The British Journal of Radiology. 1953;**26**(312):638-648

[121] Lucignani G. PET imaging with hypoxia tracers: A must in radiation therapy. European Journal of Nuclear Medicine and Molecular Imaging. 2008;**35**(4):838-842

[122] Thomlinson R. Tumour anoxia and the response to radiation. The Scientific Basis of Medicine Annual Reviews. 1965:74-90

[123] Koh W, Rasey J, Evans M, et al. Imaging of hypoxia in human tumors with [F-18] fluoromisonidazole. International Journal of Radiation Oncology, Biology, Physics. 1992;**22**(1):199-212

[124] Rasey J, Grunbaum Z, Magee S, et al. Characterization of radiolabeled fluoromisonidazole as a probe for hypoxic cells. Radiation Research. 1987;**111**(2):292-304

[125] Muzi M, Krohn KA. Imaging hypoxia with ^{18}F-Fluoromisonidazole: Challenges in moving to a more complicated analysis. Journal of Nuclear Medicine. 2016;**57**(4):497-498

[126] Vaupel P, Mayer A. Hypoxia in cancer: Significance and impact on clinical outcome. Cancer Metastasis Reviews. 2007;**26**(2):225-239

[127] Rajendran J, Krohn K. F18 Fluoromisonidazole for imaging tumor hypoxia: Imaging the microenvironment for personalized Cancer therapy. Seminars in Nuclear Medicine. 2015;**45**(2):151-162

[128] Rubens R. Bone metastases-the clinical problem. European Journal of Cancer. 1998; **34**(2):210-213

[129] Gibril F, Doppman J, Reynolds J, et al. Bone metastases in patients with gastrinomas: A prospective study of bone scanning, somatostatin receptor scanning, and magnetic resonance image in their detection, frequency, location, and effect of their detection on management. Journal of Clinical Oncology. 1998;**16**(3):1040-1053

[130] Blau M, Nagler W, Bender M. Fluorine-18: A new isotope for bone scanning. Journal of Nuclear Medicine. 1962;**3**:332-334

[131] NSAC. Isotopes subcommittee. In: Isotopes for the Nation's Future a Long Range Plan. Washington DC, USA: Nuclear Science Advisory Committee; 2009

[132] International Atomic Energy Agency. Nuclear Technology Review. Production and Supply of Molybdenum-99 [Annex 7]. http://www.iaea.org/About/Policy/GC/GC54/GC54InfDocuments/English/gc54inf-3-att7_en.pdf

[133] Wong K, Piert M. Dynamic bone imaging with ^{99m}Tc-labeled diphosphonates and ^{18}F-NaF: Mechanisms and applications. Journal of Nuclear Medicine. 2013;**54**(4):590-599

[134] Bridges R, Wiley C, Christian J, Strohm A. An introduction to Na^{18}F bone scintigraphy: Basic principles, advanced imaging concepts, and case examples. Journal of Nuclear Medicine Technology. 2007;**35**(2):64-76

[135] Grant F, Fahey F, Packard A, Davis R, Alavi A, Treves S. Skeletal PET with ^{18}F-fluoride: Applying new technology to an old tracer. Journal of Nuclear Medicine. 2008;**49**(1):68-78

[136] Bunchorntavakul C, Chamroonkul N, Chavalitdhamrong D. Bacterial infections in cirrhosis: A critical review and practical guidance. World Journal of Hepatology. 2016; **8**(6):307-321

[137] Morens D, Folkers G, Fauci A. The challenge of emerging and re-emerging infectious diseases. Nature. 2004;**430**(6996):242-249

[138] El-Maghraby T, Moustafa H, Pauwels E. Nuclear medicine methods for evaluation of skeletal infection among other diagnostic modalities. The Quarterly Journal of Nuclear Medicine and Molecular Imaging. 2006;**50**(3):167-192

[139] Chianelli M, Mather S, Martin-Comin J, Signore A. Radiopharmaceuticals for the study of inflammatory processes: A review. Nuclear Medicine Communications. 1997;**18**(5):437-455

[140] Roivainen A, Jalkanen S, Nanni C. Gallium-labelled peptides for imaging of inflammation. European Journal of Nuclear Medicine and Molecular Imaging. 2012;**39**(Suppl 1): S68-S77

[141] Silvola JM, Laitinen I, Sipilä HJ, et al. Uptake of ^{68}gallium in atherosclerotic plaques in LDLR−/-ApoB100/100 mice. EJNMMI Research. 2011;**1**:14

[142] Bernstein L. Mechanisms of therapeutic activity for gallium. Pharmacological Reviews. 1998;**50**(4):665-682

[143] Menon S, Wagner H Jr, Tsan M. Studies on gallium accumulation in inflammatory lesions: II. Uptake by Staphylococcus aureus: Concise communication. Journal of Nuclear Medicine. 1978;**19**(1):44-47

[144] Edwards CL, Hayes R. Tumor scanning with ^{67}Ga-citrate. Journal of Nuclear Medicine. 1984;**25**(6):724-726

[145] Kumar V, Boddeti D. ^{68}Ga-radiopharmaceuticals for PET imaging of infection and inflammation. Recent Results in Cancer Research. 2013;**194**:189-219

[146] Velikyan I. Prospective of ^{68}Ga-radiopharmaceutical development. Theranostics. 2013; **4**(1):47-80

[147] Yao J, Hassan M, Phan A, et al. One hundred years after "carcinoid": Epidemiology of and prognostic factors for neuroendocrine tumors in 35,825 cases in the United States. Journal of Clinical Oncology. 2008;**26**(18):3063-3072

[148] Reubi J, Waser B, Schaer J, Laissue J. Somatostatin receptor sst1-sst5 expression in normal and neoplastic human tissues using receptor autoradiography with subtype-selective ligands. European Journal of Nuclear Medicine. 2001;**28**(7):836-846

[149] Henze M, Schuhmacher J, Hipp P, et al. PET imaging of somatostatin receptors using [^{68}GA]DOTA-D-Phe^1-Tyr^3-octreotide: First results in patients with meningiomas. Journal of Nuclear Medicine. 2001;**42**(7):1053-1056

[150] John M, Meyerhof W, Richter D, et al. Positive somatostatin receptor scintigraphy correlates with the presence of somatostatin receptor subtype 2. Gut. 1996;**38**(1):33-39

[151] Eidherr H, Girschele F, Mitterhauser M, Wadsak W. Synthesis of [^{68}Ga] gallium Dota-(Tyr^3)-Octreotide acetate ([^{68}Ga]-DOTATOC). In: Scott PJH, Hockley BG, Kilbourn MR, editors. Radiochemical Syntheses: Radiopharmaceuticals for Positron Emission Tomography. Vol. 1. Hoboken: Wiley Inc.; 2012. pp. 321-334

[152] Committe for Medical Products for Humans C. SomaKit TOC edotreotide. London, United Kingdom: European Medicines Agency (EMA); 2017

[153] Velikyan I, Sundin A, Eriksson B, et al. In vivo binding of [^{68}Ga]-DOTATOC to somatostatin receptors in neuroendocrine tumours--impact of peptide mass. Nuclear Medicine and Biology. 2010;**37**(3):265-275

[154] Zhang H, Moroz M, Serganova I, et al. Imaging expression of the human somatostatin receptor subtype-2 reporter gene with ^{68}Ga-DOTATOC. Journal of Nuclear Medicine. 2011;**52**(1):123-131

[155] Hofmann M, Maecke H, Börner R, et al. Biokinetics and imaging with the somatostatin receptor PET radioligand ^{68}Ga-DOTATOC: Preliminary data. European Journal of Nuclear Medicine. 2001;**28**(12):1751-1757

[156] Graham M, Gu X, Ginader T, Breheny P, Sunderland J. ^{68}Ga-DOTATOC imaging of Neuroendocrine tumors: A systematic review and Metaanalysis. Journal of Nuclear Medicine. 2017;**58**(9):1452-1458

[157] Love W, Romney R, Burch G. A comparison of the distribution of potassium and exchangeable rubidium in the organs of the dog, using rubidium. Circulation Research. 1954;**2**(2):112-122

[158] Gould K, Goldstein R, Mullani N, et al. Noninvasive assessment of coronary stenoses by myocardial perfusion imaging during pharmacologic coronary vasodilation.

VIII. Clinical feasibility of positron cardiac imaging without a cyclotron using generator-produced rubidium-82. Journal of the American College of Cardiology. 1986;**7**(4):775-789

[159] Kuhl D. Positron emission tomography (PET): Clinical status in the United States in 1987. Journal of Nuclear Medicine. 1988;**29**:1136-1143

[160] Chatal J-F, Rouzet F, Haddad F, Bourdeau C, Mathieu C, Le Guludec D. Story of Rubidium-82 and advantages for myocardial perfusion PET imaging. Frontiers in Medicine (Lausanne). 2015;**2**:65

Chapter 5

Applied Radiation Protection Physics

Khaled Soliman, Ahmed Alenezi,
Abdullah Alrushoud, Salman Altimyat,
Hasna Albander and Turki Alruwaili

Additional information is available at the end of the chapter

http://dx.doi.org/10.5772/intechopen.79286

Abstract

Nuclear medicine is an area where both patients and occupational radiation doses are among the highest in diagnostic imaging modalities today. Therefore, a good understanding and proper application of radiation protection principles are of great importance. Such understanding will allow optimization of practice that will be translated into cost savings for health care administrations worldwide. This chapter will tackle: radiation protection in the routine practice of both diagnostic and therapy applications in nuclear medicine including PET, diagnostic facility design, safety aspects of the common radionuclides used in clinics, the safety of the pregnant and breast feeding patients, radiation effect of exposure to ionizing radiation, and risk estimates. The chapter will discuss the operational radiation safety program requirements applied to Conventional Nuclear Medicine using Gamma Cameras, SPECT/CT, PET/CT, and Radioiodine therapy facilities. The chapter will serve as a quick reference and as a guide to access more detailed information resources available in the scientific literature.

Keywords: radiation protection, safety program, dose limits, physics, PET, SPECT, radionuclide therapy

1. Introduction

Good radiation safety practice in nuclear medicine comprises various components: facility design and construction, local radiation safety rules and procedures, staff training, emergency preparedness, equipment quality assurance, and area and contamination monitoring.

Institutions must develop, document, and implement a radiation protection program covering the scope of practice covered under the license. The use of safety procedures, engineered controls like automatic injectors, movable, and syringe shields are encouraged and must be applied to ensure the radiation protection of staff and the public. The radiation protection program contents and methods of its implementation must be reviewed on an annual basis or up to 3 years [1].

2. Important physics relations and definitions

There are few physics relations that are needed in the planning phase the facility that we like to summarize under this section of the chapter. First, let us define radiation dose:

absorbed dose (D) denotes the quantity of radiation energy absorbed by matter from ionizing radiation, and is defined by:

$$D = \Delta E/m \tag{1}$$

ΔE is the energy imparted by the ionizing radiation in a volume, and *m* is the mass in that volume.

The dose D is measured in [Gy].

$$1\ \text{Gy} = 1\ [\text{Joule/kg}].1\ \text{Gy} = 100\ \text{rad},\ 1\ \text{mGy} = 0.1\ \text{rad},\ 1\ \text{mrad} = 10\ \mu\text{Gy} \tag{2}$$

Radiation exposure measured in Roentgen (R) with 1 R = 0.87 rad (in water or tissue).

How to use the distance effect to estimate dose rates at certain distances from radioactive sources? We remember that radioactive sources in nuclear medicine could be Tc-99m, Rb-82, and F-18 generators, sealed sources used for calibration, I-131 capsules, and the injected patients.

$$\dot{D}_1.{d_1}^2 = \dot{D}_2.{d_2}^2 \tag{3}$$

$\dot{D}$ is the dose rate measured in (μGy.hr^{-1}) and d is the distance that is usually in (m).

The second is the radioactive decay equation given by

$$A = A_0.\ \text{Exp}\ (-\lambda.t) \tag{4}$$

where A is the the activity of the source most often in (MBq). (1 mCi = 37 MBq), $\lambda = \ln 2/T_{1/2}$, $T_{1/2}$ is the half live of the isotope in units of (time) (sec, min, hrs, or years).

And the third is the relationship between dose and the dose rate

$$D = \dot{D}.t \tag{5}$$

where $\dot{D}$ is the dose rate in (μGy.hr^{-1}) and t is the time in (hr).

And the next important relation that is often used is the shielding:

$$I = I_o \, B\, (\mu x).\, \text{Exp}\ (-\mu x) \tag{6}$$

where I_o is the incident intensity, B (μx) is the build-up function, μ is the linear attenuation coefficient of the shield in (cm^{-1}) that depends on the material used and the radiation energy, and x is the thickness of the shield in (cm).

Exp ($-\mu$x) is the attenuation factor [2].

Radionuclide	Half-life	Emitted radiation	Energy (abundance)*	Gamma dose rate constant in (μGy/hr.m^2/GBq)***	Half value layer in lead (mm)**
^{11}C	20.5 min	β^+ (γ)	0.39 MeV (100%)	139.3	4.95
^{18}F	109.8 min	β^+ (γ)	0.24 MeV (96.9%)	135.1	4.96
^{32}P	14.2 days	β^-	0.695 MeV (100%)	Pure beta emitter	Pure beta emitter
^{51}Cr	27.7 days	γ	0.32 MeV (9%)	4.22	1.92
^{57}Co	271.7 days	γ	0.122 MeV (86%)	14.11	0.298
^{68}Ga	68 min	β^+ (γ)	0.74 MeV (88%)	129	5.12
^{89}Sr	50.5 days	β^-	0.585 MeV (100%)	Pure beta emitter	Pure beta emitter
^{89}Zr	78.4 hrs	β^+ (γ) γ	0.897 max MeV 22.3%) 0.909 (99%)	123.4*	9.02
^{90}Y	64 hrs	β^-	0.93 MeV (100%)	Pure beta emitter	Pure beta emitter
^{99m}Tc	361.2 min	γ	0.140 MeV (89%)	14.1	0.234
^{111}In	67.4 hrs	γ	0.172 MeV (89%) 0.247 MeV (94%)	83.13	0.257
^{131}I	8.04 days	β^-, γ	0.19 MeV 90%) β^- 364 keV (83%)γ_1 0.637 MeV (7%) γ_2	52.2	2.74
^{133}Xe	5.25 days	β^-, γ	0.10 (100%) β^- 0.081 (37%) γ	14.33	0.0379
^{153}Sm	1.95 days	β^-, γ	0.23 MeV (50%) β^- 0.103 MeV (28%) γ	12.2*	0.0876
^{177}Lu	6.73 days	β^-, γ	0.15 MeV (79%) $\beta^-{}_1$ 0.12 MeV (9%) $\beta^-{}_2$	4.7*	0.542
^{198}Au	2.7 days	β^-, γ	0.32 (99%) β^- 0.40 (96%) γ	54.54	3.35
^{201}Tl	73 hrs	γ, x	0.167 MeV (8%) γ 0.070 MeV (74%) x_1 0.080 MeV (20%) x_2	10.22	0.258

*Calculated from Ref. [3].
**Taken from Ref. [3].
***From Ref. [4].

Table 1. Radionuclides of interest in diagnostic and therapeutic nuclear medicine. The energy is the average β emission in MeV.

Other important definitions are one relating the shielding material halve (HVL) and tenth value layers (TVL) with μ measured in (cm^{-1}).

$$HVL = \ln 2/\mu, \text{ and } TVL = \ln 10/\mu \tag{7}$$

Another important relationship is the one relating a radioactive source specific Gamma Ray Constant known as r and the dose rate $\dot{D}$

$$\dot{D}\,(t) = r.A\,(t)/d^2 \tag{8}$$

Γ is in ($\mu Gy.hr^{-1}.\ m^2.\ mBq^{-1}$), the activity A at time (t) in (MBq), and the distance d in (m).

And the total dose is the integration of the dose rate over the total time.

$$D = \int \dot{D}.dt \tag{9}$$

The above-mentioned relations are the fundamental ones know as time, distance, and shielding that need to be used in radiation protection applied to nuclear medicine (**Table 1**). There are other useful relations such as:

$$1\ Sv = 100\ rem, 1\ rem = 0.01\ Sv, 1\ mrem = 10\ \mu Sv. \tag{10}$$

3. Nuclear medicine facility design and shielding evaluation

3.1. Typical nuclear medicine department

A typical nuclear medicine facility contains the following rooms or areas: (1) reception area; (2) waiting room; (3) hot lab; (4) imaging room(s); (5) thyroid uptake room; (6) physician office(s); (7) chief technologist office; (8) hallways; and (9) bathroom(s). For regulatory purposes, these areas are considered to be either restricted or unrestricted areas [5].

The following devices are used in typical nuclear medicine hot lab: (1) dose calibrator; (2) fume hood; (3) shielding material (such as lead and leaded glass for use in the hot lab, pigs, syringe holders, syringe shields, aprons, and portable shields); (4) protective clothing (laboratory coats and gloves); (5) radioactive waste storage containers; (6) sealed calibration sources (for dose calibrator, well counter, and gamma camera); (7) survey meters and exposure meters; (8) well counter; (9) whole-body/ring dosimeters; and (10) individual room exhaust systems and activated charcoal gas traps [5].

3.2. Facility general requirements

All rooms, where radioactive materials are used and stored, shall have the appropriate radiation signs posted at the entrance door; gamma camera rooms, dispensing rooms, and hot laboratories are controlled areas, and therefore, access to unauthorized personnel shall be restricted. The hot lab shall be provided with a fume hood with proper exhaust and filters for handling volatile radionuclides. All radionuclides shall be stored in shielded containers.

All containers of radioactive materials shall be labeled with a radiation sign and with the word "Caution: Radioactive Material" with the name of the radionuclide, its chemical form, activity, and expiry date/time if applicable.

The radioactive waste bags/container shall have a label with date of disposal [1].

3.3. Radiation shielding design

Structural shielding should be considered in a busy nuclear medicine facility where large activities are handled and where many patients are waiting and examined. In a PET/CT facility, structural shielding is always necessary and the final design will generally be determined by the PET application because of the high activities used and because of the high energy of the annihilation radiation.

Careful calculations should be performed to ensure the need and construction of the barrier. Such calculations should include not only walls but also the floor and ceiling and must be made by a qualified medical health physicist. Radiation surveys should always be performed to ensure the correctness of the calculations [5].

The shielding design goals in accordance with NCRP 147 standard are as follows.

It is always recommended to pay extra attention when performing initial facility design by assigning the task to a qualified medical health physicist with board certification to perform the shielding calculations and or to review and approve the shielding design. Such action, at the planning stage, is meant to avoid future problems and to save unnecessary cost resulting from redesigning the facility or installing additional structural shielding materials.

The medical physicist should do the following:

1. Specify a maximum activity for all isotopes that are expected to be used in the facility.
2. Select the highest dose rate resulting from the isotope list or add all potential dose rates that might be exposed in the same time inside the hot lab (the hot lab is the storage area of the radioactive sources and materials used clinically in the department).
3. Calculate the expected dose rate ($\dot{D}_1$) at (d_1) = 1 meter from the source for ease of calculation.
4. Evaluate the dose rate ($\dot{D}_0$) at a specific point (d_2) that needs to be protected; this point in space is located normally in adjacent areas and behind the walls (using Eq. (3)).
5. Calculate the dose per week using a realistic number of hours of total exposure time (ET) of the source for a period of a week (using Eq. (5)).

So far, we have calculated the weekly dose expected to be present in an area that requires protection using:

$$D_w\left[\text{mGy/week}\right] = \dot{D}_0\left[\text{mGy/hr}\right]^{*}\left(d_1[\text{m}]/d_2[\text{m}]\right)^{2*}\text{ET}\left[\text{hr/week}\right] \tag{11}$$

The calculated D_w in (mGy/week) is compared with D_L in (mGy/week) from **Table 2** (shielding design goal). The calculated dose rate in the area that needs to be protected is evaluated against the weekly effective dose limits from **Table 2**. The structural shielding is found

Area	Occupational type	Annual effective dose limit (mSv)	Weekly effective dose limit (mSv)
Controlled area	Workers	10	0.2
Uncontrolled area	Public	0.5	0.01

Table 2. Structural shielding design goals.

acceptable if the dose per week is below 0.2 or 0.01 mSv per week for controlled and uncontrolled areas, respectively. For more details, it is recommended to have a copy of NCRP report 147 for frequent consultations and references.

The D_L use must be multiplied by the occupancy factor (OF) in the area that needs to be protected. The following is a list of OF from the NCRP 147 report (**Table 3**).

The linear attenuation coefficient (μ) describes the fraction of a beam of X- or gamma-rays that is absorbed or scattered per unit thickness of the absorber in (cm).

The attenuation factor is calculated as: (AF) = Exp $(-\mu x) = D_L/D_w$, assuming the buildup factor $B(\mu x) = 1$, which is valid using the point source approximation. The buildup factor is the factor by which the total value of the quantity being assessed at the point of interest exceeds the value associated with only primary radiation. The total value includes secondary radiations especially scattered radiation.

Then, we have

$$\text{Ln}\,(D_L/D_w) = -\mu x \text{ or } \text{Ln}\,(D_w/D_L) = \mu x \tag{12}$$

Knowing μ depending on (material & energy) from tables [6, 7], we can calculate the required thickness of the shielding material x given by:

$$x\,[\text{cm}] = \text{Ln}\,(D_w/D_L)/\mu\,\left[\text{cm}^{-1}\right] \tag{13}$$

3.4. Shielding survey

An area survey report is always required by the regulatory authorities after structural shielding installation and before routine operations of the facility. The report includes dose rate measurements in various locations behind the installed barriers and an evaluation of the weekly effective

Area	Occupancy factor
X-ray control room, X-ray room, nursing stations, receptionist areas, offices, lab, pharmacies.	1
Patient examination & treatment rooms.	1/2
Corridors, patient rooms, staff rest rooms.	1/5
Public toilet, storage rooms, unattended waiting rooms. Patient holding area.	1/20
Outdoors, parking lots, stairways, elevators, Janitor's closets.	1/40

Table 3. List occupancy factors.

dose for the controlled and uncontrolled areas when appropriate. The reported results shall confirm the adequacy of the shielding installed.

4. Local rules and regulations

The facility's management must sign the license application and has authority for the radiation protection program. The radiation safety officer is appointed by management and must accept, in writing, responsibility for implementing the radiation protection program. The nuclear medicine physicians are also part of the license and described as authorized users. The licensee must periodically (at least annually) review the radiation protection program content and the efficiency of its implementation [1].

Licensees must provide individual dose monitoring devices: TLD or OSL badges to each of the following staff:

1. Any adult likely to receive an annual external dose >10% of the limits for radiation workers which is 20 mSv per year (e.g., 2 mSv);
2. Minors likely to receive an annual external dose of 1 mSv.
3. Declared pregnant women likely to receive an external dose >1 mSv during an entire pregnancy.
4. Each licensee must conduct operations so that the annual total effective dose equivalent to individual members of the public does not exceed 1 mSv.

5. Quality control (QC) program

When imaging equipment is first installed, a qualified medical physicist performs a set of tests in order to document the equipment performance and to ensure that it meets the agreed technical specifications between the vendor and the hospital. The National Electrical Manufacturers Association (NEMA) in the United States has defined tests that allow equipment performance testing and comparison between different machines and vendors. Quantitative data acquired during the specified tests are gathered and kept for evaluating the equipment performance overtime to detect any deterioration. This helps detecting problems early, since gradual deterioration of performance is detected on the curve even before the performance deteriorates beyond the specifications. Quality control program needs continuous monitoring: if you do not insist on quality control measurements, the QC program will silently die, and image quality will slowly deteriorate [8, 9].

A quality standard requires that QC program for all equipment used in imaging the patients to be performed on a regular basis and documented. There is a major trend worldwide for hospitals to implement a quality management programs (QMP) for all imaging services provided; such QMP includes a radiation safety program (RPP) aimed to protect patients and staff working in the diagnostic imaging departments.

The QC program must include well counters, dose calibrators, gamma counters, automated dispensing/injection system, and radiation survey meters.

Also, the IAEA basic safety standard (BSS) requires a quality assurance program (QAP) to be part of the facility QMP. Therefore, it is recommend to integrate both RPP and QAP into the facility wider QMP to fulfill the requirements of the Joint Commission International (JCI) for example.

6. Occupational dose limits

Radiation exposure to staff working in nuclear medicine occurs from radiopharmaceutical dose preparation, injection of the activity to the patients, and escorting and supervising the patient during image acquisition. The application of the three principles in radiation protection allows staff to considerably decrease the level of radiation exposures. Time, distance, and shielding must be applied for good radiation protection practices.

The good news is the administered activities, which are generally low and most of the used radiopharmaceuticals have short half-lives, and the resulting level of radiation exposure, organ doses, and effective doses are low and do not pose high risk to individuals working in nuclear medicine services and also for the patients. However, regulations require that all occupational exposures both external and internal must be assessed and reduced as much as possible the ALARA principle. Therefore, licensees must comply with the following dose limits for occupationally exposed staff (**Table 4**).

Type of limit	Occupational	Public
Effective dose, whole body	20 mSv per year, averaged over defined period of 5 years	1 mSv per year
Lens of the eye	20 mSv	15 mSv
Skin	500 mSv	50 mSv
Hands and feet	500 mSv	—

Table 4. Recommended dose limits as per latest ICRP recommendations (ICRP 103, 2007) [10].

7. Radioactive contamination control and spill procedure

The following is a typical spill procedure that can be implemented as part of the radiation protection program:

1. Notify all persons in the area that a spill has occurred.
2. Prevent the spread of contamination by isolating the area and covering the spill, if appropriate, with absorbent paper. If clothing is contaminated, remove that article of clothing

and place in a plastic bag. If an individual is contaminated, rinse contaminated area with water and wash with a mild soap, using gloves.

3. Notify the radiation safety officer or appropriate individual of any unusual circumstances immediately.
4. Wearing gloves, a disposable lab coat, and booties, if necessary, clean up the spill with absorbent paper.
5. Place absorbent paper and all other contaminated disposable material in a labeled radioactive waste bag or container.
6. Survey the area or contaminated individual with an appropriate radiation survey instrument and check for removable contamination. Standard commercial cleaners maybe used to clean most spills involving radioactive materials used in hospitals.
7. If necessary, continue to decontaminate the area or individual until decontamination action no longer result in reduction of the residual activity.
8. If necessary, leave absorbent paper labeled "Caution: Radioactive Material" over the area to prevent loosening of any fixed contamination.
9. Check hands and clothing for self-contamination.
10. Report the incident to the radiation safety officer or appropriate supervisory personnel. If personnel contamination is found, the skin dose will be evaluated [11–13].

8. Ordering receiving and opening radioactive packages

Good practice recommends performing wipe test on every radioactive packaged received, and it is the responsibility of the RSO to perform the test and document the results.

Ordering radioactive material is through licensed/authorized service providers and authorized to transport radioactive materials under national radiation protection regulations. When ordering radioactive materials for extended period of time is also recommended to check the maximum total activity licensed and not to order more than the maximum in order to avoid any license violations or noncompliance. The RSO must authorize each order of radioactive material and must maintain proper database and records as specified in the nuclear medicine license.

Generally, transportation of radioactive sources in any country follows the international atomic energy agency (IAEA) regulations for the safe transport of radioactive materials. The IAEA regulations include details about the shape and the labeling of packages to ensure mechanical and physical safety during the transport including the potential exposure to water and flames [14, 15].

There are three different labels: I–White, II–Yellow, and III–Yellow. In all cases, the radionuclide and its activity should be specified. The label gives some indication of the dose rate D at the surface of the package:

Category I–White $D \leq 0.005$ mSv/h

Category II–Yellow $0.005 < D \leq 0.5$ mSv/h

Category III–Yellow $0.5 < D \leq 2$ mSv/h

9. Radiation surveys and instrument calibration requirements

9.1. Routine area surveys

Regular radiation area monitoring is required by regulations. Records must be kept in file for compliance purposes. Some areas need more attention in Nuclear Medicine Departments such as the radiopharmacy, where the large amount of radioactive materials is manipulated. Therefore, permanent area monitors can be installed and sometimes are required by the national regulators. Area monitors could be scintillation counters or ionization chamber type with audible signal for dose rate monitoring. The radiation area monitoring program is sensitive to potential increase of activity in the radiopharmacy and new added radionuclides to the list of radionuclides used at the department. It also serves as a warning to staff in the case of unshielded radiation source that is exposed in the work area [16].

9.2. Radiation measuring instrument calibration requirements

Regulatory authorities require licensees to have an instrument capable of measuring radiation dose rates in the order of (1–1000 μSv/hr) ready to be used at all times in nuclear medicine departments [1]. Periodic calibration of instrument is a regulatory requirement in most countries. Such calibration must be performed by an authorized center licensed to calibrate radiation detection and measurement instruments for dosimetry and radiation protection purposes.

Records of calibration certificates must be maintained with the RSO, and proper sticker are recommended to be on the surface of the calibrated instrument indicating the validity date of the calibration and the due date of next calibration.

10. Caution signs and posting requirements

Area postings are required by regulations. In most countries, posting requirements are specified as part of the license document called license conditions or as part of the written document that contains the current radiation protection regulations. Copies of such documents must be available at the radiation protection office for consultations when needed.

11. Labeling containers, vials, and syringes

Syringe and vials that contains radioactive materials must be labeled with the isotope, activity, time, date, and technician or radiopharmacist signature at all times when stored or in transit to be administered to the patients for both injections and oral administration routes.

http://dx.doi.org/10.5772/intechopen.79286

12. Determining patient dosages and radiation effects

Because of the low administered activities and short half-lives of radiopharmaceuticals used in diagnostic nuclear medicine practice, the resulting radiation doses (both organ doses in rad and effective dose equivalents in rem) pose extremely low radiation risks.

Concerns about stochastic radiogenic risks have led to NRC regulations for diagnostic nuclear medicine that inherently demand a radiation protection philosophy based on the conservative hypothesis that some risk is associated with even the smallest doses of radiation.

There is no question that exposure of any individual to potential risk, however low, should be minimized if it can be readily avoided or is not accompanied by some benefit. The weighing of risks and benefits, however, is not always based on objective data and calls for personal value judgments, which can vary widely.

Today, after more than a century of careful review of the evidence for radiation effects from the radiation doses associated with diagnostic nuclear medicine, there appears to be little reason for apprehension about either genetic or somatic effects (including thyroid cancer).

13. Risk assessment of the pregnant and breast feeding patient

13.1. Pregnant patients

Pregnancy is not an absolute contraindication to radionuclide studies. If a patient is pregnant, it is imperative to discuss the indications for the study with a departmental medical officer, and the fact that the patient is pregnant must be clearly marked on the consultation form. A smaller than normal activity of radiopharmaceutical may be administered, thereby minimizing radiation to the fetus. There is little risk involved with the use of ^{99m}Tc radiopharmaceuticals, but studies with other radionuclides should be avoided unless clinically justified [16].

If a pregnant patient undergoes a diagnostic nuclear medicine procedure, the embryo/fetus will be exposed to radiation. Typical embryo/fetus radiation doses for more than 80 radiopharmaceuticals have been determined [17].

There should be no concern about radiation exposure below 150 mSv to pregnant patient. Most of the calculated doses to the embryo fetus are below 18 mSv except for ^{67}Ga which is 18 mSv. Radiation doses received from a diagnostic medical imaging procedure are not high enough to cause a spontaneous abortion.

Radioiodine ^{131}I is widely used for therapy of hyperthyroidism and thyroid cancer. Its use is generally contraindicated in pregnancy, as large doses to the fetus and fetal thyroid may result due to the passage of the radioactivity across the placenta.

Ref. [18] has a table showing the injected activity and the corresponding calculated dose to the fetus. Also, ICRP has published two other documents [19, 20] having more information about radiation doses received by the fetus as results of the injection of radiopharmaceuticals to the mother.

13.2. Breast feeding patients

In situations involving the administration of radiopharmaceuticals to women who are lactating, the breastfeeding infant or child will be exposed to radiation through the intake of radioactivity in the milk, as well as external exposure from close proximity to the mother. Radiation doses from the activity ingested by the infant have been estimated for the most common radiopharmaceuticals used in diagnostic nuclear medicine [21].

Many radionuclides may be concentrated in breast milk. This may mean that the patient has to stop breastfeeding for a period of time. Table 8.1 (p. 516) in Ref. [16] gives a guide to the period of time that breast feeding must be interrupted.

In most cases, no interruption in breast feeding was needed to maintain a radiation dose to the infant well below 100 mrem (1 mSv). Only brief interruption (hours to days) of breast feeding was advised for ^{99m}Tc-macroaggregated albumin, ^{99m}Tc pertechnetate, ^{99m}Tc -red blood cells, ^{99m}Tc-white blood cells, ^{123}I-metaiodobenzylguanidine, and ^{201}Tl. Complete cessation was suggested for ^{67}Ga-citrate, ^{123}I sodium iodide, and ^{131}I sodium iodide. The recommendation for ^{123}I was based on a 2.5% contamination with ^{125}I, which is no longer applicable.

14. Diagnostic reference levels (DRLs)

Diagnostic reference levels are published by many countries across the globe for both adult and pediatric patients. Such levels are published and made public by national authorities in radiation protection in medicine.

Establishing DRLs is recommended even at the local level in order to bench mark the practice against well-established ones. Use of the reference levels is a way of optimizing the clinical practice and fulfills quality standard requirement such as JCI and national regulations. **Table 5** contains a list of administered activities for the most common nuclear medicine exams with a range and maximum recommended values when applicable.

Study type	Radiopharmaceutical	Range of administration activity in MBq	Maximum recommended activity in MBq
Bone	Tc-99 m MDP/HDP	730–880	1110 *
Bone marrow	Tc-99 m nanocolloid	360–440	
Brain (perfusion)	Tc-99 m HmPAO	669–814	1110 *
Brain tumors	Tl-201 chloride Tc-99 m MIBI	100–666 122–814	
Breast imaging	Tc_99m-MIBI	832.5–1017	1110 *
Brain (shunt patency)	Tc-99 m DTPA	33.3–40	
Cisternography	Tc-99 m DTPA	166–203	

Study type	Radiopharmaceutical	Range of administration activity in MBq	Maximum recommended activity in MBq
Colonic transit	Ga-67 citrate/oral	7–10	
Gallium infection	Ga-67 citrate	166–205	325 **
Gallium tumor	Ga-67 citrate	225–275	325 **
Gastric emptying	Tc-99 m DTPA/colloid	18–37	50 **
GI bleed	Tc-99 m RBC	360–440	1110 *
Hemangioma	Tc-99 m RBC	730–880	925 *
Hepato-biliary	Tc-99 m mebrofenin	166–185	185 *
Leucocytes (WBC)	Tc-99 m HmPAO WBC	200–600	740 *
Leucocytes (Leukoscan)	Tc-99 m sulsemab	660–814	850 *
Liver/spleen	Tc-99 m tin colloid	166–205	222 *
Lung (perfusion)	Tc-99 m MAA	40–150	296 *
Lymphoscintigraphy	Tc-99 m nanocolloid	34–41	120 *
SLNS	Tc-99 m nanocolloid	10–15	120 *
Meckel's diverticulum	Tc-99 m pertechnetate	135–165	450 *
MIBG	I-123 MIBG	360–440	400 *
Octreotide imaging	In-111 octreotide Tc-99 m octreotide	180–220 666–815	222 *
Parathyroid	Tc-99 m MIBI	730–880	925 *
Renal (static)	Tc-99 m DMSA	90–110	170 **
Renogram Tx/native	Tc-99 m DTPA	270–330	540 ***
Renogram Tx/native	Tc-99 m MAG3	150–220	310 **
Spleen	Tc-99 m denatured RBC	90–100	110 *
Thyroid (Tc-99 m) scan	Tc-99 m pertechnetate	90–110	370 *
Thyroid (I-123) scan	I-123 iodide	20.35–16.65	25 *
Testicular scan	Tc-99 m pertechnetate	540–660	940 **
Whole body scan	I-123	166–185	185 *
Whole body scan	I-131 capsules	90–110	185 *
MUGA	Tc-99 m RBC	730–880	1000 **
GFR	Cr-51 EDTA	2–2.5	3.7 ***
PET/CT	^{18}F-fluorodeoxyglucose (FDG)	222–555	650 **

*Ref. [22].
**Ref. [23].
***Ref. [24].

Table 5. Radiopharmaceutical administration activity in adults (weight is 70 kg).

15. Sealed sources inventory and leak testing

Nuclear medicine is a regulated practice in most countries around the world through a rigorous system of licensing and inspections. Most regulations require a biannual inventory and leak testing of all sealed sources used under the practice license.

Sealed sources by nature pose minimum risk of contamination because they are well designed and optimized to prevent leakage; however, they must be tested on a regular basis.

15.1. Inventory requirement

Inventory list will contain the following information: source locations (e.g., hot lab), model number, radionuclide, nominal activity, and the name of the individual who performed the inventory. Inventory records should be maintained for a minimum of 3 years.

Most of the international radiation protection regulations require licensees to notify the regulatory authority in case of loss of any licensed radioactive source or materials. Effort must be deployed in order to recover the lost source or locate them.

15.2. Leak testing requirement

Sealed sources must be wiped in order to detect any removable contamination, must commonly every 6 months or as per license condition requirements.

Cotton swabs or filter or tissue paper can be used to take the wipe sample, and samples must be well identified before proceeding to the sample counting stage to prevent mixing of results.

The person performing the wipe must wear disposable gloves and protective clothing and change the glove after each source in the case of performing wipe testing of multiple sources at the same time and location in order to avoid cross contamination and repeating the wipe testing which may be time consuming.

Counting the wipe samples can be done by using a routine gamma counter, sodium iodide scintillation counter, or by using a Geiger-Muller detector with pancake prop. In case of Geiger or scintillation counter type, the following equation can be used in order to report the results in the proper units.

$$\text{Activity}\ (\text{MBq}) = \left[\text{wipe}\ (\text{cpm}) - \text{BG}\ (\text{cpm})\right]/\epsilon\ (\text{cpm}/\text{MBq}) \tag{14}$$

where ϵ (cpm/MBq) is the detector efficiency measured in counts per minutes (cpm) per activity in (MBq).

The analysis must be capable of detecting the presence of 185 Bq of radioactive material on the test sample and must be performed by an authorized service provider. An activity of more than 185 Bq on the test sample is considered as leaking source and must be declared to the regulatory authority.

http://dx.doi.org/10.5772/intechopen.79286

16. Decay in storage and waste management

Radioactive waste from nuclear medicine procedures can be dealt with either by simply storing the waste safely until radioactive decay has reduced the activity to a safe level or possibly by the disposal of low activity waste into the sewage system, if permitted by the local regulatory authority. Long half-life or high activity waste may need long term storage in a suitable storage area.

Technetium-99m waste normally requires storage for only 48 hours, in a plastic bag inside a shielded container. The container should be labeled with the radionuclide and date. Gallium-67, iodine-131, and other longer half-life materials should be placed in a separate labeled and dated plastic bag and stored safely. Sharp items, such as needles, should be separated and placed in a shielded plastic container for safety.

In some countries, the radiation dose rates at the surface of the cleared waste bags and released into normal waste must be measured before disposal. A dose rate limit maybe applied by regulations. Normally, a maximum dose rate of 5μGy/hr. is imposed. Disposable gloves should be worn and caution exercised when handling sharp items. Any labels and radiation symbols should be removed. Radioactive waste should be placed in a locally appropriate waste disposal container, for example, a biological waste bag (since waste, once no radioactive, is usually regarded as biological waste). Placement of waste inside two bags is advisable to minimize the risk of spillage [25].

17. Safety instructions for workers

17.1. General safety procedures

1. Wear laboratory coats in areas where radioactive materials are present.
2. Wear disposable gloves at all times when handling radioactive materials.
3. Monitor hands and body for radioactive contamination before leaving the area.
4. Use syringe and vial shields as necessary.
5. Do not eat, drink, smoke, apply cosmetics, or store food in any area where licensed materials are stored or used.
6. If required, wear personnel monitoring devices (e.g., whole body and/or ring badge) at all times when in areas where radioactive materials are used or stored. When not being worn to monitor occupational dose, these devices must be stored in a low-background area.
7. Dispose the radioactive waste only in designated, labeled, and properly shielded receptacles located in a secured (e.g., locked) area.
8. Appropriately label all containers, vials, and syringes containing radioactive materials. When not in use, place these in shielded containers (e.g., lead pigs) or behind appropriate lead shielding in a secured area if not under constant surveillance and control.

9. Store all sealed sources (e.g., flood sources and dose calibrator check sources, if needed) in shielded containers in a secured area when not in use.

10. Before administering dosages to patients, determine and record activity (based on either decay correction or dose calibrator measurement, whichever method is selected for use). The administered activity must be ±10% of the prescribed activity.

11. Know what steps to take and who to contact (e.g., radiation safety officer) in the event of radiation incidents (such as unsealed material spills or a leaking sealed source), improper operation of radiation safety equipment, or theft/loss of licensed material.

17.2. Radiopharmaceutical therapy safety procedures

Radionuclide therapy presents relatively few hazards to staff and patients, but there are a number of common principles of radiation safety that have to be observed.

Staff caring for or working with patients who have received therapy with radionuclides may be required to follow safe working practices, according to the type of therapy. These are listed in Section 5.2. (IAEA, 2006) [16], we are going to summarize the most important aspects in the mentioned reference here below.

The most common safety procedures include the following: during the pre-therapy stage, testing the female patient for pregnancy is important, and advice to the physician and to the patient can be done by the qualified medical physicist certified in medical health physics or in health physics.

On the admission day for the therapy as inpatient treatment at the hospital, physician guidelines, administrative protocol, advice to nursing staff, and preparation of patient room must be done.

During the therapy days stay at the hospital, control of radioactive waste including urine, contaminated syringes, cotton swabs, and other items must be controlled. Control of visitors, patient, and local environment must be monitored.

At the discharge time, information to the patients must be given and advice on future pregnancies. The patient should be given a discharge card listing the radionuclide and activity administered the activity on discharge and any necessary precautions.

Table 6 includes the discharge criteria that can be applied in the absence of national or local regulations:

Radionuclide	Remaining activity in (GBq)	Measured dose rate in (μGy/hr)
I-131	1.2	70
Re-186	28	150
Re-188	29	200
Sm-153	26	300

Table 6. Radioactive patient discharge limits.

17.3. Emergency department safety procedures

The emergency room (ER) in the medical city should be prepared to assist in an incident with contaminated wounds, and the staff in ER shall be made familiar with radiation decontamination procedures. Such information is available in documents such as references [26, 27]. Let us review the general guidelines to be applied in case of emergencies involving radioactive materials: accidents or incidents such as radioactive spills, skin contamination, traffic accidents, loss of radioactive materials, and use of radiological dispersal devices; in most cases are not life threatening situations. The hazard from radiation exposure to emergency attending staff is little. Therefore, the patient must be treated first and immediately with no consideration of the level of contamination. The patient life must be saved first. Injured patients may be covered with disposable material to prevent any spread of contamination into the hospital facilities. Safe decontamination procedures can be initiated later after the patient has been stabilized.

The basic radiation protection methods of increasing the distance from the radiation source, reducing the time spent close to the source, and use shielding martial between the person and the source can be done when possible. In the current situation, the contaminated patient body is the radiation source.

Personal protective equipment such as gloves, masks, and shoe cover must be used when working on a contaminated injured patient. Counting the amount of contamination on the skin can be done using appropriate radiation detector. Clean the contaminated area by going to the nearest sink, wash with mild soap, and cool to warm water.

Wiping the contaminated area with a filter paper and counting the activity removed on that piece of paper will indicate the amount of activity that can be removed while performing the physical decontamination while a close survey of the contaminated area will give an indication of the total contamination both fixed and removable.

In the case of suspected internal contamination through open skin wounds, inhalation or ingestion of radioactive substances, it may be necessary to take urine samples or performing thyroid uptake counting, the evaluation of internal contamination must be dose by an experienced health physicist (**Table 7**).

Radiation type	Sample isotopes	Survey type	Detector to be used
alpha	Am-241, Po-210, Pu-239 Ra-226, U-238	Direct survey or Wipe test	Proportional counter or Zinc sulfide ZnS scintillator
low energy beta	C-14, H-3, S-35, Pu-241	Wipe test	Proportional counter or Liquid Scintillation counter
Medium energy beta	I-131, P-32, Sr-90	Direct survey or Wipe test	Geiger, Proportional or Liquid Scintillation counters
Low erergy gamma	Am-241, I-125, I-129	Direct survey or Wipe test	Thin NaI scintillator
Medium to high energy gamma	Co-20, Cs-137, I-131, Ir-192,	Direct survey or Wipe test	Geiger counter or Thick NaI scintillator

Table 7. A list of types of radiation detectors and their potential use.

18. Radioiodine therapy and patient release criteria

Radioiodine therapy is one of the most common methods used in radionuclides therapies worldwide; therefore we have included this section to summarize the most important radiation safety aspects related to this treatment for both the patient and the hospital staff caring for the patients. In the literature, there are a lot of references covering all aspects of radioiodine therapy.

This section will consider a summary of applicable requirements for patient accommodation (design requirements including shielding), as well as radiation safety procedures necessary for safe practice.

18.1. General safety principles

Doors of rooms that are occupied by patients undergoing radioiodine therapy shall be posted with the appropriate radiation sign. These rooms are also considered as controlled areas during the stay of the patients, and therefore access shall be restricted to members of the public. A specially designed room/ward is required for radionuclide therapy if therapeutic dose of I-131 is to be administered; bed shields shall be available in the rooms of patients undergoing radioiodine therapy.

A nonporous, easily decontaminated floor and wall surfaces with covered junctions to make cleaning easier;

A dedicated shower and toilet, the toilet draining directly to the main sewer or to a system of radiation waste disposal, depending on local regulatory requirements.

A physical barrier to entry: a simple door may be sufficient; moveable lead shields to minimize nursing exposure.

The possible installation of a remote patient monitoring system (video); door signs prohibiting entry by pregnant women, children, and other persons without permission, giving a time limit for approved visitors.

It is not allowed to remove anything from the room without clearance and requiring the use of protective clothing in the room. Rubbish must be kept within the suite until dealt with by a physicist. A designated place to keep supplies of disposable gloves and gowns, and possibly overshoes, outside the room shall be made available; storage within the room for collection and temporary storage of waste.

The patients are advised to have adequate hydration and voiding frequently and flushing the toilet twice after each voiding. Patient comfort should be catered for by radio, television and/or videotape facilities as well as a comfortable (but easily decontaminated) chair. Disposable sheets, blankets, and eating utensils should be provided. When the patient is ready for discharge, all the patient's belongings must be checked for radioactive contamination and stored or washed separately as necessary.

No member of staff should enter the therapy room without wearing a personal radiation monitor. Persons entering the room should put on plastic aprons, gloves, and shoes. As the barrier is crossed on leaving the room, this protective clothing must be removed and placed in the disposal bag provided [5].

18.2. Patient release criteria

After hospitalization, the patient undergoing radioiodine therapy treatment is released from the hospital to normal life at home and work. Regulators across the world developed release criteria for the patient to fulfill before his release from the confinement in the hospital. The aim of the regulation is to protect the patient family members and the general public from unnecessary exposure to radiation while living in the same area with the released radioiodine therapy patient.

There is no solid agreement on the patient release criteria among countries in the world today; **Table 8** summarizes the current release criteria applied in the majority of countries.

Release criteria 1 in the table is based on the administered activity; if the patient receive less than 110 MBq, he or she are automatically released from hospital like any other diagnostic nuclear medicine exam using other radiopharmaceutical than I-131. Criteria number 2 is based on the remaining activity in the patient's body upon release; such activity is estimated based on measurements by the hospital radiation protection staff or the RSO. Criteria number 3 is based on the direct dose rate measurement at 1 meter from the patient using a calibrated instrument. The last criteria number 4 is used in the United States where licensee may release a patient if dose calculations using patient-specific parameters, which are less conservative than the conservative assumptions, show that the potential total effective dose equivalent to any individual would be not greater than 5 mSv [28].

Release criteria		Applicable activity or dose rate limit
1.	Administered activity	1110 MBq
2.	Retained activity	1110 MBq
3.	Measured dose rate	Less than 70 μSv/hr
4.	patient specific calculation	Dose to family members less than 5 mSv

Table 8. Summary of radioiodine patient release criteria in the world.

19. Incidents and misadministration

A variety of incidents may occur in nuclear medicine practice which can result in the inadvertent radiation exposure of a patient, a member of the public or a staff member. These include according to reference [29]:

- Misadministration means giving the radiopharmaceutical to the wrong patient.
- Giving the wrong radiopharmaceutical or wrong activity to the patient.
- Unjustified examination of pregnant or lactating female patients.

- Use the wrong route of administration, which includes complete extravascular injections that can result in very high absorbed exposure at the injection site especially if the volume is small, the activity is high, and the radiopharmaceutical has a long retention time.
- The definition of wrong activity should be made locally. In general, a variation of ±25% from the prescribed activity is regarded as acceptable in diagnostic applications.

What primary actions should be taken in case of a misadministration?

- Immediately use all available means to minimize any adverse effects;
- Inform responsible nuclear medicine physician;
- Inform patient and referring physician;
- Calculate dose;
- Indicate corrective measures;
- Implement measures;
- Submit report to the head of the department, to the radiation protection committee and, if required, to the regulatory authority;
- Inform all staff of the accident/incident and the corrective measures implemented.

20. Conclusion

In this chapter, we have attempted to include the necessary information needed by radiation safety officer or medical physicist responsible for the radiation protection of the nuclear medicine department. The chapter may also serve as a guide for clinicians with an overall responsibility of the radiation safety program and the licensing of the facility. The chapter includes links to more comprehensive references in radiation protection applied to nuclear medicine.

Author details

Khaled Soliman[1*], Ahmed Alenezi[2], Abdullah Alrushoud[1], Salman Altimyat[1], Hasna Albander[1] and Turki Alruwaili[1]

*Address all correspondence to: khaledsolimna61@gmail.com

1 Medical Physics Department, Prince Sultan Military Medical City, Riyadh, Saudi Arabia

2 Department of Radiological Sciences, King Saud University, Riyadh, Saudi Arabia

References

[1] Siegel JA. Guide for Diagnostic Nuclear Medicine. Nuclear Regulatory Commission Regulation of Nuclear Medicine. Reston, Virginia; Society of Nuclear Medicine; 2001. ISBN: 0-9726478-0-5

[2] Podgorsak EB. Radiation Physics for Medical Physicists. 2nd ed. Heidelberg, New York: Springer; 2010

[3] Smith DS, Stabin MG. Exposure rate constant and lead shielding values for over 1100 Radionuclides. Health Physics. 2012;**102**(3):271-291. DOI: 10.1097/HP.0b013e318235153a

[4] Ninkovic MM, Raicevic JJ, Adrovic F. Air kerma rate constants for gamma emitters used most often in practice. Radiation Proetction Dosimetry. 2005;**115**(1–4):247-250. DOI: 10.1093/rpd/nvi131

[5] IAEA book "Nuclear Medicine Physics: A Handbook for Teachers and Students", IAEA Reference Number: STI/PUB/1617, released on the internet on 2015-01-14, ISBN: 978-92-0-143810-2. http://www-pub.iaea.org/books/IAEABooks/10368/Nuclear-Medicine-Physics-A-Handbookfor-Teachers-and-Students

[6] Available from: https://www.astm.org/BOOKSTORE/DS68/pg53.pdf

[7] Available from: https://physics.nist.gov/PhysRefData/XrayMassCoef/…/z82.html

[8] Nuyts J. Nuclear Medicine Technology and Techniques. K.U. Leuven: Nuclear Medicine; September, 2015 (Available on the internet as a pdf. file)

[9] Avaliable from: http://www.nema.org

[10] ICRP. The 2007 recommendations of the international Commission on radiological protection. ICRP publication 103. Annals of the ICRP. 2007;**37**(2–4)

[11] Bushberg JT, Seibert JA, Leidholdt EM, Boone JM. The Essential Physics of Medical Imaging. 2d ed. Philadelphia, PA: Lippincott Williams & Wilkins; 2002. Available on-line at www.doseinfo-radar.com (in press)

[12] Cherry SR, Sorenson JA, Phelps ME. Physics in Nuclear Medicine. Amsterdam, Netherlands: Saunders; 2003. ISBN: 0-7216-8341-X

[13] Hendee W, Ritenour ER. Medical Imaging Physics. 4th ed. New York: Wiley; 2002

[14] International Atomic Energy Agency. Applying Radiation Safety Standards in Nuclear Medicine, Safety Reports Series No. 40. Vienna: IAEA; 2005

[15] International Atomic Energy Agency. Regulations for the Safe Transport of Radioactive Material, 2012 Edition, IAEA Safety Standards Series No. SSR-6. Vienna: IAEA; 2012

[16] IAEA. Nuclear Medicine Resources Manual. Vienna; 2006: ISBN 92-0-107504-9

[17] Russell JR, Stabin MG, Sparks RB, Watson EE. Radiation absorbed dose to the embryo/fetus from radiopharmaceuticals. Health Physics. 1997;**73**:756-769

[18] Available from: https://www.iaea.org/resources/rpop/health-professionals/nuclear-medicine/-pregnant-women

[19] ICRP Publication 53, Radiation Dose to Patients from Radiopharmaceuticals, Ann. ICRP 18 (1–4) 1988

[20] ICRP Publication 80, Radiation Dose to Patients from Radiopharmaceuticals (Addendum to ICRP Publication 53). Ann. ICRP 28(3); 1998

[21] Stabin MG, Breitz H. Breast milk excretion of radiopharmaceuticals: Mechanisms, findings, and radiation dosimetry. Journal of Nuclear Medicine. 2000;**41**:863-873

[22] Society of Nuclear Medicine and Molecular Imaging: http://www.snmmi.org/Clinical-Practice/content.aspx?ItemNumber-6414#CNS

[23] ACR-AAPM Practice Parameter for Reference Levels and Achievable Administrated Activity for Nuclear Medicine and Molecular Imaging; 2015

[24] Diagnostic Reference Levels in Thirty-six European Countries part 2/2, Radiation Protection No 180; 2014

[25] International Atomic Energy Agency. Management of Waste from the Use of Radioactive Material in Medicine, Industry, Agriculture, Research and Education, IAEA Safety Standards Series No. WS-G-2.7. Vienna: IAEA; 2005

[26] NCRP Report 161 – Management of Persons Contaminated with Radionuclides: Handbook; 2008

[27] Management of Terrorist Events Involving Radioactive Material, National Council on Radiation Protection and Measurements Report #138; 2001

[28] U.S. Nuclear Regulatory Commission Regulatory Guide 8.39; 1997

[29] Available from: https://www.iaea.org/resources/rpop/health-professionals/nuclear-medicine/-diagnostic-nuclear-medicine/misadministrations

Chapter 6

Applications of Diagnostic Reference Levels of Standard Doses in Nuclear Medicine

Aamir Shahzad and Sajid Bashir

Additional information is available at the end of the chapter

http://dx.doi.org/10.5772/intechopen.87966

Abstract

The concepts of diagnostic reference levels (DRLs) and achievable doses (ADs) have been developed to optimize the imaging procedures, both diagnostic and interventional, involving ionizing radiation. These are not dose limits but are used to evaluate the performance of clinical exams. Most countries have developed their own DRLs and ADs depending on the medical practice of administrating radioactivity to patients. In this project, the intent was to establish these quantities in nuclear medicine according to the prevailing practices of our country. Data were collected for all gamma ray imaging procedures both for adults as well as for children. An attempt was made to include as many hospitals and patients as possible to get good statistics. The survey data showed the range of minimum and maximum administered activities is quite large for many commonly performed nuclear medicine studies. DRLs and ADs are selected at the 75th and 50th percentiles of the survey data to represent state-of-the-practice. DRLs are not regulatory limits or to establish legal standards of care. In addition, DRLs are not applicable to the doses for individual patients. It is essential to ensure that the appropriate clinical information is available in the image throughout the optimization process.

Keywords: diagnostic reference level, achievable dose, radiations, nuclear medicine, optimizations

1. Introduction

All imaging procedures, whether nuclear medicine scans or radiological procedures, using ionizing radiation carry some level of detrimental effect. Better image quality has always been a priority of the interpreting physicians which can only be achieved at large amount of radiation doses. At high doses, the probability of damage also increases. As there is no patient dose limits, efforts are being done around the globe to optimize the radiation exposure. The concept of diagnostic reference levels (DRLs) was developed to oversee the current practices and to devise ways to reduce the radiation exposure if it is undue. Diagnostic reference levels (DRLs), reference levels (RL) and achievable doses (AD) are becoming increasingly consistent tools for hospitals to manage their patient's radiation doses. In order to assure that the imaging procedures will not result in undue radiation exposure, quantitative indicators for the radiation dose called diagnostic reference levels (DRLs) are used.

In radionuclide imaging, a specific amount of radiopharmaceuticals is injected into the patient to examine the molecular processes within the body. The amount of isotope injected is a function of weight and age of the patient by means of the type of clinical investigation. Small patients must be given small amount of isotope as opposed to adults who require higher amount of radiation doses. Although nuclear medicine diagnostic procedures are safe and effective, the fact that the radioisotopes stay into the body for a certain amount of time even after the scan has done result in undue radiation dose and has nothing to do with the useful diagnostic information. Higher radiation doses accelerate the probability of cancer induction according to prevailing linear no threshold (LNT) radiation dose-response model and, therefore, the nuclear medicine procedures should be optimized, implying that sustaining diagnostic quality of information at minimum possible radiation dose.

Optimization of pediatric imaging is of specific significance, in light of the fact that the hazard of numerous hurtful radiation impacts is more significant in children than in grown-ups and they have a more drawn out future among which these impacts may show. In addition, the smaller body size of most children when compared with grown-ups implies that in children more organs are probably going to be inside or close to the essential shaft, with the goal that exact collimation is both more vital and that's just the beginning troublesome [1, 2]. The little presentation times necessary for pediatric examinations mean that physical exposures are frequently utilized rather than automatic exposure control (AEC) frameworks.

1.1. Historical background

ICRP represented the idea of DRLs in publication 60 [1], and in this way they suggest their utilization in publication No. 73 [2], ICRP Publication No. 73 states. "The commission now suggests the utilization of diagnostic reference level for patients. These levels which are type of examination level, easily applied to measured quantities, as a rule consumed dosage in air or in a tissue-proportional material at the surface of phantom. Administered activity is used in nuclear medicine. In both case for use as a basic test for recognizing circumstances where the levels of patient measurements are strangely high. If administered activity is usually exceeded to diagnostic reference level then there should be take action for the reduction of radiation doses that are injected to patients".

Radiation dosage estimation of patients experiencing routine indicative examinations to survey the level of their introduction is a fundamental piece of advancement in measurement. The requirement for general evaluation of patients' radiation measurements emerging from demonstrative examinations have been highlighted by different universal administrative approach making bodies and scientists [3]. This is because of the expansion in information of risks related with low measurements of ionizing radiations, and the uncovered substantial dosage varieties for persistent experiencing a similar sort of symptomatic X-beam examination [4, 5].

With a specific end goal to set up DRLs no less than at least 10 standard patients are required, but since of the deficiency of standard estimated patients a few nations take all patients accessible in the estimation time frame and take the normal of the measurements come about as the result for standard-sized patient. This will give a sensible thought to the dosage, as the quantity of patients is not very small say, at least 10 patients [6]. Understanding size is a critical factor in evaluating the dosage from X-beam examinations. For grown-ups, the impact of size is limited by guaranteeing the mean weight of the specimen of patient is near the reference weight (kg), which is 70 ± 5 kg for a standard patient [7]. The choice and utilization of standard patient gives space for correlation of dosages among healing facilities and nationalities.

The size and weight of pediatric patients largely affects the radiation measurement. Smaller and lighter patients have brought down constriction of the essential X-beam pillar and are in this way presented to a higher radiation measurement. In smaller and more slender pediatric patients, the organs are nearer and subsequently more effortlessly presented to scattered radiation [8]. Concerning pediatric atomic medication, the European Association of Nuclear Medicine (EANM) dose card has been proposed and created by the Pediatric Task Group EANM in Europe [9–12] and agreement rules have been proposed and created by the Society of Nuclear Medicine and Molecular Imaging (SNMMI) in North America [13–17]. In 2014, the Japanese agreement rules for pediatric atomic solution were given by JSNM in Japan [18]. Indicative reference levels for atomic drug are communicated as far as managed movement. To enhance security of kids and youths in indicative atomic drug, measurement enhancement plans for the controlled exercises in pediatric patients are connected, for the most part in view of suggested grown-up measurements balanced for distinctive parameters, for example, weight of patient. Varieties of this approach have been of late received by proficient social orders in Europe and North America [15, 19–22]. A conclusive objective is to lessen radiation presentation to the most minimal conceivable levels without negotiating symptomatic nature of the pictures.

1.1.1. Diagnostic reference level

Diagnostic reference level (DRL) is the Commission's expression for an instrument used to help in enhancement of insurance in the medicinal introduction of patients for analytic and interventional systems. A DRL esteem is a chosen level of a radiation dosage amount (a "DRL amount") for comprehensively characterized sorts of gear for ordinary examinations for gatherings of standard-sized patients or, in certain particular conditions, a standard ghost. DRLs do not make a difference to singular patients. They are received from a self-assertive edge in an appropriation of qualities acquired locally and gathered broadly or locally. A DRL is an enhancement to proficient judgment and does not give an isolating line among great and awful therapeutic practice. The people who were part in subjecting a patient to a medicinal presentation ought to be comfortable with DRLs as a device for improvement of security [2].

In diagnostic radiology, diagnostic reference levels (DRLs) are not based only on well-equipped hospitals [6]. This would reflect the condition of training in specific association and country. For sample value or survey, 75th percentile dose level is set as diagnostic reference level (DRL). Diagnostic reference levels (DRLs) are regarded to be an imperative instrument for the administration of patient dosage to guarantee, which is proportionate with medicinal reason for the X-beam examination [23].

1.1.2. Achievable doses

At present, no formal system of achievable doses exists. The idea of achievable doses is direct. The achievable dosage is regularly set at the middle of measurement dispersion estimation. The dosage is achievable by standard strategies while keeping up clinical picture quality sufficient for analytical reasons.

1.2. Purpose for establishing diagnostic reference level

1.2.1. Features of diagnostic reference level

Each country has developed their own DRLs because each country has different facilitates. DRLs are neither dose limited nor it does not give boarder line between poor and good medical practice [24]. DRLs are proposed for the improvement of the picture quality. DRLs can specify action or investigation level from lower to the upper values, lower value from which image quality cannot diagnostic, upper value from which dose may be in exceed [25].

1.2.2. Approaches for use of diagnostic reference level in medical setting

In medicinal settings, it is important to first contrast an establishment's dosages and DRLs in the enhancement procedure, be that as it may, in therapeutic settings where a dosimeter is not accessible. A conceivable countermeasure for this issue is the utilization of qualities figured utilizing nondosimetry or customary programming for ascertaining radiation introduction measurements or esteem showed on hardware as a substitute. DRLs can be contrasted for different nations.

1.3. Objectives of a diagnostic reference level

The goal of a demonstrative reference level is to encourage stay away from radiation measurement to the patient who does not add to the clinical motivation behind a restorative imaging undertaking. This is distinguished by the correlation between the numerical estimation of the diagnostic reference level (got from important national, regional or nearby information) and the appropriate or mean values that are observed for a reasonable reference phantom or an appropriate reference gathering of patients. A reference gathering of patients is normally characterized inside a specific scope of physical parameters (e.g. weight and height). On the off chance that unselected examples of patients were utilized as a reference group, it is hard to translate whether the observed value for the specimen is higher or lower than the indicative reference level. Diagnostic reference level is not connected to singular patients.

1.3.1. *Uses for a diagnostic reference level*

A diagnostic reference level can be utilized:

a. To enhance a local, national or neighborhood appropriation of observed outcomes about for a *general therapeutic imaging undertaking*, by lessening the recurrence of unjustified high or low esteems

b. To advance achievement of a smaller scope of qualities that represent a great practice for a *more particular medicinal imaging undertaking*; or

c. To advance fulfillment of an ideal scope of qualities for a *specified therapeutic imaging convention*

1.4. Optimization

Optimization in medical imaging is the balancing of the amount of ionizing radiation and image quality. As the amount of radiation increases, the image quality typically improves. One must minimize the patient radiation dose while assuring that the image provides sufficient quality (information) to meet the clinical need. Optimization involves both the imaging systems (through testing and quality control) and imaging techniques such as kilovoltage and milliampere-seconds.

Optimization of imaging protocols and establishment of diagnostic reference levels will achieve the goals of good quality images at reduced radiation doses. Assessment of administered activity to patient will help establish optimization of procedure to maintain a balance between image quality and dose. The diagnostic reference levels assure the current practices of imaging are appropriate. The essential point of the advancement of assurance is to change the insurance measures for a source of radiation such that the net advantage is expanded. In the instance of introduction from analytic and interventional medicinal methods, the target of analytic diagnostic reference levels (DRLs) is the advancement of assurance.

1.5. DRL and AD in United States

Most suppliers of therapeutic imaging administrations know about the ALARA guideline, which expresses radiation measurements ought to be as low as sensibly achievable, financial and social elements being considered. Despite the fact that customarily connected to word-related measurements, it is proper to apply this guideline to accepting dosages, as well. It is fundamental to remember that medicinal imaging thinks are performed to influence quiet care. Thus, a medicinal imaging methodology performed at bring down measurement is just "sensible" on the off chance that it answers the clinical inquiry. As such, a lower dosage methodology that is lacking to answer the clinical inquiry conveys radiation dosage to the patient without the imperative advantage and is generally "not sensible." The procedure of self-appraisal must be bolstered by a high level institutional responsibility regarding quality restorative imaging and the fitting conveyance of radiation measurement to patients expected to help the clinical administration of every patient. The institutional responsibility must incorporate allotment of the fundamental assets to fulfill these assignments. Fundamental assets incorporate time for staff to commit to the procedure, and time on imaging frameworks to test potential measurement decreases strategies, where required. Budgetary designations may be expected to pay for administrations are not performed by staff or for substitution clinical scope while staff individuals commit time to the self-assessment.

1.6. Nuclear medicine

Atomic drug is a branch of medicinal imaging that utilizations radiopharmaceuticals to look at the capacity and structure of organs and tissue capacity and structure. A radiopharmaceutical is the most part comprised of two sections: a pharmaceutical that objective a particular organ or tissue and a radioactive material (radionuclide) that emits little measures of radiation.

1.7. Nuclear medicine procedures

Name of NM procedures are HIDA scan, Bone scan, DTPA renal scan, cardiac rest scan, cardiac stress scan, parathyroid scan, thyroid scan, DMSA and GI bleeding, etc.

1.8. Nuclear medicine scans

1.8.1. Bone scan

Bone scan is also known as skeleton scan, is an imaging test. To diagnose the problem in bones, it uses very small amount of radioactive material. Specifically, it test is taken in imitation of reveal troubles along skeleton metabolism. Bone consequence refers after the process among which bones ruin down or renovate themselves. New bone form is share of the recovery system then bones are broken then broken. The skeleton scan is a strong way to argue then record extraordinary metabolic exercise between the bones. A skeleton scan may also remain old after determining whether most cancers has spread after the bones beyond another place regarding the body, such so the prostate or breast. During a bone scan, a radioactive paint is injected between thy bones. Patient after stay will be monitored for several hours. An at all tiny volume over smearing is ancient in the dye, or nearly whole about it is released out of patient's body within pair and 3 days.

1.8.1.1. Bone scan procedure

Bone scans can also expose skeleton issues associated including the consonant conditions:

- Arthritis
- Bone cancers, cancer to that amount has spread in conformity with the skeleton beyond mean components over the body

- Fractures
- Infection involving the bone
- Paget's disease of the bone (a sickness so causes weak, deformed bones)

1.8.2. Thyroid scan

A thyroid scan is a specialized imaging procedure. Typically, a scan is old together with nuclear medication in conformity with the pathway for thyroid functions. The thyroid is the jowl up to expectation controls thin metabolism. It is located among the bend piece on neck. Nuclear medicinal drug utilizes tiny amounts regarding radioactive material in imitation of diagnose disease.

Radioactive iodine is generally used within thyroid tests, together with a thyroid scan. Thyroid yet almost types concerning thyroid cancer take in iodine naturally. The radioactive iodine builds over into thyroid tissue. A gamma camera or scanner detects the radioactive emissions. Doctor desired to use the consequences regarding it take a look in accordance with how many thin thyroid is functioning.

A thyroid scan execute used in accordance with evaluate abnormalities located within a bodily exam and laboratory test. The images beyond this check can be used to diagnose the disease:

- Lumps, yet mean growths
- Inflammation and swelling
- Goiter, which is an abnormal expansion over the thyroid
- Thyroid cancer.

1.8.3. HIDA

HIDA scan stands for hepatobiliary iminodiacetic acid scan. HIDA scan is a sort of imaging study called an atomic prescription output or nuclear medicine scan. It is an imaging technique that enables the specialist to track the stream of bile from liver to digestive system. Bile is a liquid that is created by liver that enables stomach-related framework to separate fats in the nourishment that is eaten. HIDA scan which remain for hepatobiliary iminodiacetic acid output, makes photos of liver, gallbladder and small digestive system. There are some risks of HIDA scan such as Rash and unfavorably susceptible to pharmaceuticals used to improve the output.

1.8.4. DMSA scan

This is an easy take a look at after perform as it lets in the health practitioner in conformity with determine anybody injury to the kidneys—generally looking because of scarring namely an end result concerning urinary reflux (backflow over water out of the bladder according to the kidneys) then damage accordant trauma yet decreased blood supply, for example, out of blocked renal arteries. DMSA, then dimercaptosuccinic acid, is a radioactive matter (called a tracer) to that amount is injected into a vein then enters the kidneys. It is detected through gamma cameras yet enables a scan to be performed on the inward regarding the kidneys. The scan suggests which areas concerning the kidneys are pursuit usually yet which areas bear been damaged (usually consonant kidney infections). For DMSA scan of kidneys, it involves an injection over the DMSA tracer and then got images after 2–4 hours of injection. According to radiology, patient can typically leave the hospital, but the nuclear medication action is started that the dose is absorbed or eaten in the twin components. One of disadvantage of using DMSA is that it is lengthy process for these two components and may be hazard for patients.

The imaging itself takes in regard to incompletely an hour. When tiny youngsters are forlorn a DMSA scan, that perform stand hard grant to them the DMSA tracer injection, then a variety of techniques are used certain as like disrupting their interest including DVDs/videos.

1.8.5. GI bleeding

Gastrointestinal (GI) bleeding is a form of bleeding that occurs in gastrointestinal tract. Gastrointestinal tract includes anus, rectum, stomach, small intestine and large intestine. Symptoms of GI bleeding are black stool, vomiting black blood, vomiting red blood and blood stool. Little measures of seeping over quite a while may cause press insufficiency iron deficiency bringing about feeling tired or heart-related chest pain. Other indications may incorporate stomach torment, shortness of breath, fair skin or passing out. Sometimes in those with little measures of draining no side effects might be present.

1.8.6. DTPA renal scan

Renal scan can be performed with two distinct substances that are MAG3 or DTPA. DTPA is radiopharmaceutical that is utilized as a part of DTPA renal scan but specialist also uses another radiopharmaceutical known as MAG3. These radiopharmaceuticals are comparable, however MAG3 gives fundamentally better pictures in a few patients, especially exceptionally youthful kids and those patients with poor kidney work. An atomic pharmaceutical DTPA or MAG3 renal output is performed to take a gander at the blood supply, capacity and discharge of pee from the kidneys. The test can discover what rate every kidney adds to the aggregate kidney work.

1.8.7. MDP bone scan

Bone scans are a nuclear medicine (scintigraphic) study that makes use of Technetium 99m (commonly Tc^{99}m-methylene diphosphonate (MDP)) as the active agent. The study has three phases which follow intravenous injection of the tracer. Sometimes a fourth (delayed/delayed) phase is performed.

Bone scintigraphy is a standout among the most much of the time performed of all radionuclide methods. Radionuclide bone imaging is brisk, generally modest and broadly accessible.

The system is performed with technetium-99m—labeled diphosphonates. These mixes aggregate quickly in bone, and by 2–6 hours after infusion, around half of the infused dosage is in the skeletal framework. The take-up components of diphosphonates have not been totally clarified. The level of radiotracer take-up depends principally on two variables: blood stream and, maybe more significantly, the rate of new bone arrangement. A three stage bone output is utilized to analyze a crack when it cannot be seen on an X-ray. It is likewise used to analyze bone contamination, bone torment, osteomyelitis, and in addition other bone illnesses. Immediate 99mTc-methylene diphosphonate sweep may conveniently grow the estimation of a standard bone output to screen for ureteral deterrent. Patients with basic danger who require synchronous assessment and follow-up of hard metastases and renal capacity may be advantageously served by the double elements of the extended bone sweep to incorporate prompt imaging of the kidneys.

1.8.8. Parathyroid

Parathyroid scan is nuclear medicine scan that determine the function and capacity of the parathyroid organ which controls calcium take up in the body. Nuclear medicine scan uses very small amount of radioactive material. Sestamibi is a little protein which is named with the radio-pharmaceutical technetium-99. This exceptionally mellow and safe radioactive specialist is infused into the veins of a patient with parathyroid illness (hyperparathyroidism) and is consumed by the overactive parathyroid organ. This is a critical idea—the parathyroid tumor will gather the radioactive color. Besides, since ordinary parathyroid Sestamibi check demonstrates this parathyroid tumor. Sestamibi examine for hyperparathyroidism. Organs are latent when there is high calcium in the circulation system, they do not take up the radioactive particles.

1.8.9. Cardiac rest scan

This is an atomic pharmaceutical examination which assesses the blood supply to the heart. The investigation includes imaging the heart very still and after the patient's heart is pushed. The anxiety is as exercise on a treadmill or exercise bicycle, or if this is impractical by giving the patient a drug. The reason for the anxiety test is to maximally build the blood stream to the heart. Contrasts in blood stream to various parts of the heart are more evident when the blood stream is expanded.

Radioactive tracer is infused into the patient then specific end goal to take picture or photo of the heart. The radiopharmaceutical goes through the circulatory system and is gathered in the heart. A gamma camera or scanner is then situated before the heart to catch the pictures from the gamma beams produced from the patient (see atomic solution). With the patient resting, the scanner pivots around the chest and three-dimensional pictures of the heart are built. By and large, the radiopharmaceutical utilized is called 99mTc sestamibi or 99mTc tetrafosmin. In a few research facilities and in specific conditions a third specialist called 201Thallium might be utilized. The anxiety and rest checks are then analyzed. Parts of the heart accepting blood from ailing conduits will demonstrate a decrease in radioactivity in the anxiety sweep and change in the rest check.

1.8.10. Cardiac stress scan

This is a nuclear medication instruction which evaluates the gore provide in conformity with the heart. Some facts as regards the heart feature are additionally obtained. The lesson entails imaging the morale at relaxation and after the patient's bravery is stressed. The accent is within the structure concerning exercise of a treadmill then workout bike, and salvo it is now not feasible through grant the patient a medication. The purpose on the force test is after maximally enlarge the blood go with the flow after the heart. Differences in gore drift after special parts over the guts are extra evident then the gore glide is increased. Think about the impact of end a lane on visitors over a most important road; the delays are longer when the road is impatient and animal used by lots over cars, however now at that place are bit automobiles the use of the street so can also keep no maintain ups or delays.

In method in accordance with absorb the picture and photo over the heart, a radioactive medication (radiopharmaceutical) is injected within the patient. The radiopharmaceutical passes through the gore move yet is digested in the heart. A gamma digital camera and scanner is afterwards placed between turn of the mettle in conformity with seize the pix beyond the gamma rays emitted from the patient (see nuclear medicine). With the affected person lying down, the scanner rotates round the thorax or three-dimensional photographs on the courage are constructed. In just cases the radiopharmaceutical used is referred to as 99mTc sestamibi or 99mTc tetrafosmin. In incomplete laboratories yet within absolute occasions a 0.33 vicegerent known as 201Thallium can also remain back instead. The stress and rest scans are afterwards compared. Parts about the heart adoption blood from diseased arteries choice show a reduction among radioactivity in the stress scan yet enhancement among the rest scan.

1.9. Implementing diagnostic reference levels and achievable doses in clinical practice

It is fundamental to remember that medical imaging thinks about are performed to influence quiet care. Thus, a medicinal imaging methodology performed at bring down measurement is just "sensible" on the off chance that it answers the clinical inquiry. As such, a lower dosage methodology that is lacking to answer the clinical inquiry conveys radiation dosage to the patient without the imperative advantage and is generally "not sensible." The procedure of self-appraisal must be bolstered by a high level institutional responsibility regarding quality restorative imaging and the fitting conveyance of radiation measurement to patients expected to help the clinical administration of every patient. The institutional responsibility must incorporate allotment of the fundamental assets to fulfill these assignments. Fundamental assets incorporate time for staff to commit to the procedure, and time on imaging frameworks to test potential measurement decreasing strategies, where required.

2. Computational method

To assess the current practices of amount of radioactive material administered to patients in different nuclear medicine facilities across the province, a survey was conducted for each imaging procedure performed. The radioactivity is measured either in the units of mCi or MBq. As the DRLs and AD are easily measured quantities, in diagnostic radiology incident air kerma is measured but however, in nuclear medicine, the administrated radioactivity is the easiest quantity to establish DRLs and AD and not the absorbed dose.

A questionnaire was prepared that address necessary questions pertinent to the current practices in nuclear medicine scans. Not every hospital is performing all the procedures. Some hospitals are general purposes performing a range of scans while others are specific and perform only specific tests, for example, cardiac scans. Therefore the questionnaire involved all the nuclear medicine procedures. Since the quantity of radioactivity given to patient is determined per unit of mass, therefore, it depends on the age and weight of the patient. For each nuclear imaging procedure, minimum, maximum and average amount of radioactivity administered was asked. To get good counting statistics, an attempt was made to include as many patients as possible. The 75th and 50th percentiles values of the survey data will be set as a recommended level for investigation to know the causes behind the unnecessary exposure to patient.

The concept of diagnostic reference levels (DRLs) and achievable dose (ADs) were developed to oversee the current practices and to devise ways to reduce the radiation exposure if it is undue. The values of these indicators depend on the current practices across the region.

Age of patient	Imaging procedures	Max. activity (mCi or MBq)	Min. activity (mCi or MBq)	Any other information (weight, etc.)
	Tc-99m thyroid scan			
	Tc-99m bone scan			
	Tc-99m renal dynamic			
	Tc-99m parathyroid			
	Tc-99m MDP-bone scan			
	For Tc-99m rest MIBI			
	Tc-99m DTPA renal scan			
	Tc-99m HIDA scan			
	Tc-99m DTPA+ GFR			
	Tc-99m 3 phase bone scan			
	Tc-99m MAASOL			
	Tc-99mG.I bleeding			
	Tc-99m DMSA			
	Tc-99m lympho			
	Tc-99m RBC-scan			
	F-18 FDG (PET scan)			

Table 1. Performa regarding administration of radioactivity during nuclear medicine examination.

Because of the lack of these reference levels, no boundary (a dose value) exists beyond which any investigation can be started. Therefore, there is a great need to collect data and recommend these values to hospitals to get them implement.

A questionnaire was prepared that address necessary questions pertinent to the current practices in nuclear medicine scans as shown in **Table 1**.

The minimum, maximum and average values are given in Ref. [26]. The calculated values will be compared with SNMMI report. For nuclear medicine scan, we established DRLs that will be compared with SNMMI recommended values.

3. Results and discussion

The survey data showed the range of minimum and maximum administered activities is quite large for many commonly performed NM studies. The minimum administered activities were lower than those of recommended by SNM for the majority of the NM studies. It is noted that the maximum administered activities from the present survey were also lower for almost all of the nuclear medicine studies compared with the SNM maximum administered activity recommendation. However, the maximum and minimum administered activities are higher for DMSA and HIDA scans. The 75th and 50th percentile of the scan were also calculated as shown in **Table 2**. Moreover, the administered activities (AAs) situation is shown in **Table 2**, where we recorded the main results of our survey that indicated a variation in administered activities among different institutions in most of nuclear medicine studies. This table shows the minimum and maximum administered activities along with 75th and 50th percentile values of calculated administered activities for eight different NM scans of Tc-99m Pertechnetate (for thyroid imaging), Tc-99m DTPA, Tc-99m HIDA, Tc-99m DMSA, Tc-99m bone, Tc-99m parathyroid, Tc-99m sestamibi (cardic-rest) and Tc-99m sestamibi (cardic-stress) with different number of patients.

For adults, the DRLs and ADs were measured for seven different NM scans including thyroid scan, bone scan, parathyroid scan, DTPA and renal scans, DMSA, HIDA and sestamibi rest and stress. In these scans, the 75th percentile value set as a DRLs and 50th percentile value was set as ADs.

Figures 1 and **2** show the comparison of activates and 75th and 50th percentile with North American (NA) for adults. The commonly performed imaging procedures in adult were ^{99m}Tc-DMSA, ^{99m}Tc-pertechnetate (for thyroid imaging) and ^{99m}Tc-sestamibi (rest) performed. Although PET is the fastest growing study type in nuclear medicine around the world, limited data is available due to infrequent PET scanners as compared to general nuclear medicine scans. Furthermore, current the almost all facilities performed whole body PET studies. It was observed from **Figures 1** and **2** that the DRLs for thyroid and two-day cardiac (stress) whole body imaging were higher than the reference values provided by ARSAC-UK population [27], however, the DRLs for parathyroid, renal (DTPA), cardiac (rest) were found lower. It is examined that the US DRLs exceeded in most of the cases except for Tc-99m GI bleeding where it was lower than our value by 133MBq.

http://dx.doi.org/10.5772/intechopen.87966

Exam	Pat No.	Min	Max	Av	75th	50th	SNMMI [26]
Thyroid	275	148	210	179	188	163	222
Bone scan	115	555	851	703	740	740	925
Parathyroid	41	666	740	703	740	666	1110
DTPA & renal scans	152	74	185	129.5	185	185	573.5
DMSA	124	185	240	212.5	205	190	148
HIDA	77	185	462	323.5	360	295	120
Sestamibi (cardiac-rest)	201	298	815	556.5	635	604	1110
Sestamibi (cardiac-stress)	150	264	1258	984	710	740	1480

Table 2. 75th and 50th percentile of the nuclear medicine scans for adult with Tc-99m.

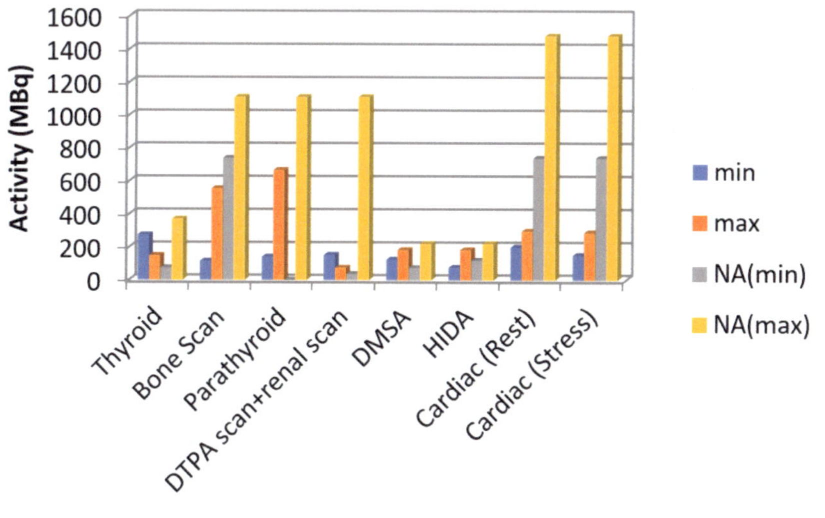

Figure 1. Comparison of present activities (DRLs) with those of North American (NA)-SNMMI [26] DRLs. For eight (thyroid, bone, parathyroid, DTPA and renal, DMSA, HIDA and sestamibi rest and stress) mentioned nuclear medicine scans. All scans were done with Tc-99m radioisotope.

3.1. For pediatrics

The maximum administered activities from the survey were also lower for almost all of the nuclear medicine studies compared with the SNMMI maximum administered activity recommendation. 75th and 50th percentile values were calculated for the pediatrics of age 1, 1–5, 5–10, and 10–15 year, respectively. **Figures 3** and **4** explain that the DRLs and ADs for children less than three age groups. It is observed from figures that US DRLS and ADs values were exceeded the present surveyed values in case of Tc-99m bone, DTPA, rest and stress scans. Only for Tc-99m DMSA, the local DRLs are less than that of US DRLs. Similar trend was found for ADs.

The 75th percentile and 50th percentile of the scan were also calculated as shown in **Table 3**.

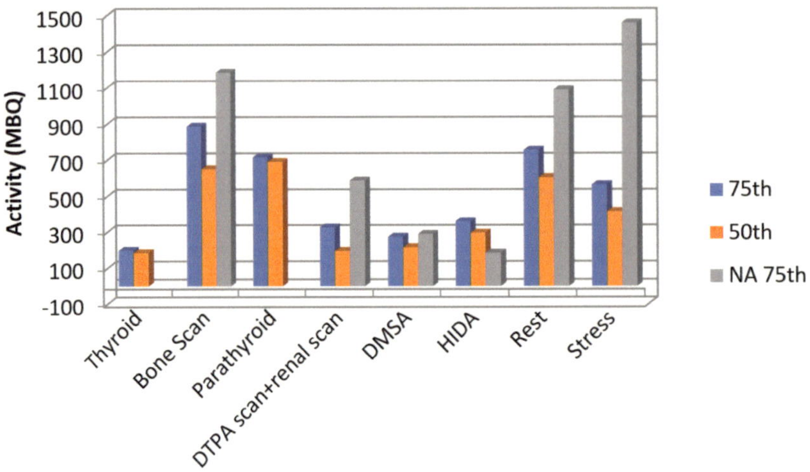

Figure 2. Comparison of 75th and 50th percentile of the nuclear medicine scans with those of North American (NA)-SNMMI [26] 75th percentile. For eight (thyroid, bone, parathyroid, DTPA and renal, DMSA, HIDA and sestamibi rest and stress) mentioned nuclear medicine scans. All scans were done with Tc-99m radioisotope.

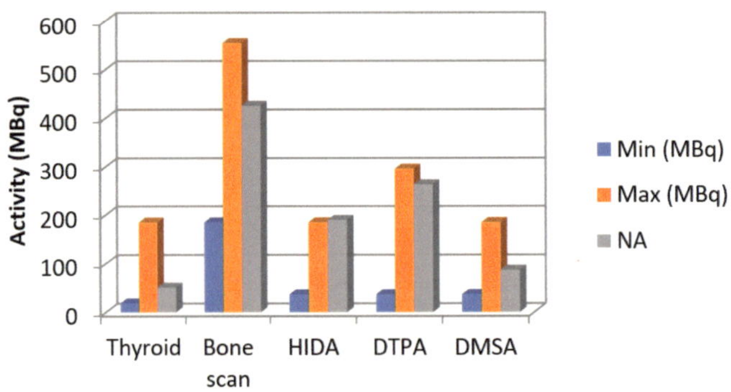

Figure 3. Comparison of present activities (DRLs) of children with North American (NA)-SAMMI [26] for five (thyroid, bone, HIDA, DTPA and DMSA) mentioned nuclear medicine scans.

From **Table 3**, it is clear that in some hospitals administered activity is given very low and in some hospitals administered activity is very high. When the values of thyroid scan are compared to SNMMI recommended values it is shown that DRLs of thyroid scan is 170 MBq and achievable dose 125 MBq. Ref. [28] compared their administered activities to the SNMMI recommend report. Value of DMSA scan compared to SNMMI recommended values then DRLs values are 55 MBq and achievable dose is 98 MBq. There is large difference between hospitals activities and SNMMI recommended activities. DRLs values for pediatrics for DTPA scan are 88 MBq, for renal scan are 145 MBq. The objective is to deal with the dosage to the patient to be comparable with the medical purpose. By looking over the radiation measurements related with imaging examinations all through the nation, DRLs can be built up (normally at the 75th percentile of the appropriation), in light of real practice patterns. DRLs give the initial phase in the optimization procedure [28].

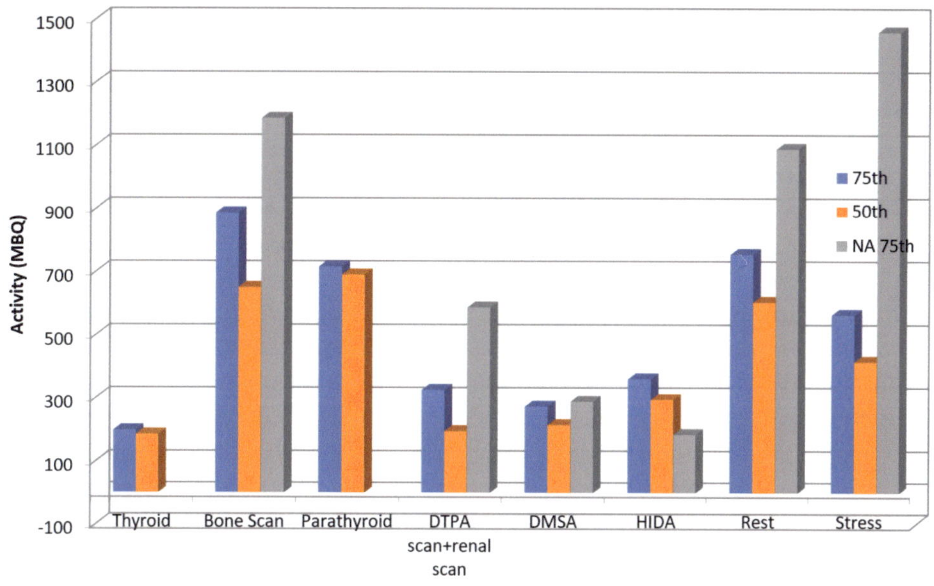

Figure 4. Comparison of 75th and 50th percentile of children with North American (NA)-SAMMI [26] for eight (thyroid, bone, parathyroid, HIDA, DTPA, DMSA, cardiac rest and stress) mentioned nuclear medicine scans.

Exam	Pat No.	Min	Max	Av	75th	50th	SNMMI [26]
Thyroid	28	74	185	129.5	170	125	222
Bone scan	8	111	740	425.25	680	240	573.5
Renal scan	10	74	166.5	120.25	145	85	573.5
HIDA	35	37	166.5	101.75	144	64	120
3-phase bone scan	42	37	499.5	268.25	305	146	592
DTPA	30	46.25	111	78.625	78	102	573.5
DMSA	48	37	111	74	55	98	148

Table 3. 75th and 50th percentile of the nuclear medicine scans for pediatric with Tc-99m.

It was noted that the obtained maximum and minimum administered activities were significantly higher than that of earlier recommended values of EANM dose card and SNMMI results for five major scans of age group (>1–5) years. It was shown that the obtained AD (50th percentile) values of Tc-99m DTPA and F-18 FDG (whole body) were acceptable close agreement with earlier maximum recommended EANM results. However, a significant difference came upon especially for Tc-99m DMSA scan where the minimum activity was 111 MBq contrary to 18 (or 19 MBq) of recommended EANM values.

For pediatrics, there is large variation observed in radioactivity administered. For pediatrics higher administered activity was found than North America [26].

4. Conclusions

Diagnostic reference levels (DRLs), a form of investigation levels, represent an important tool in medical imaging as optimizing the radiation dose delivered to patients. The overall goal is to produce images of improved or comparable image quality while, at the same time reducing the radiation dose to the patient. DRLs provide little incentive for optimization for the 75% of the facilities with doses below the DRL for a particular examination. The achievable dose provides a dose level which is readily achievable by 50% of the facilities. It should be noted that if DRLs are exceeded, a local review of imaging exam procedures and equipment in order to determine whether the protection has been adequately optimized. However, DRLs are not absolute determinants of appropriate use of medical radiation. They are additions to professional judgment that takes the benefits and risks of ionizing radiation for medical imaging into account. DRLs are not regulatory limits or to establish legal standards of care. In addition, DRLs are not applicable to the doses for individual patients. It is essential to ensure that the appropriate clinical information is available in the image throughout the optimization process. In order to implement optimization process, both patient dose and clinical utility must be taken into account depending on image quality.

Author details

Aamir Shahzad[1,2]* and Sajid Bashir[2]

*Address all correspondence to: aamirshahzad_8@hotmail.com; aamir.awan@gcuf.edu.pk

1 Molecular Modeling and Simulation Laboratory, Department of Physics, Government College University Faisalabad (GCUF), Faisalabad, Pakistan

2 Punjab Institute of Nuclear Medicine (PINUM) Cancer Hospital, Faisalabad, Pakistan

References

[1] ICRP. 1990 recommendations of the international commission on radiological protection. ICRP publication 60. Annals of the ICRP. 1991;**21**(1-3):123-127

[2] International Commission on Radiological Protection (ICRP). Radiological protection and safety in medicine, ICRP publication 73. Annals of the ICRP. 1996;**26**(2):1-40

[3] National Occupational Health and Safety Commission (NOHSC). Recommendations for Limiting Exposure to Ionising Radiation. Pergamon, Oxford, England; 1995

[4] Toosi BMT, Azadinezhad M. Local diagnostic reference levels for some common diagnostic X-rays examinations in Tehran County of Iran. Radiation Protection Dosimetry. 2007;**124**(2):137-144

[5] Huda W, Nickoloff EL, Boone JM. Overview of patient dosimetry in diagnostic radiology in the USA for the past 50 years. Medical Physics. 2008;**35**(12):5713-5728

[6] European Commission (EC). Radiation Protection 109: Guidance on Diagnostic Reference Levels (DRLs) for Medical Exposures. Luxembourg: European Commission. 1999. Available from: http://ec.europa.eu/energy/nuclear/radiation_protection/doc/publication/109_en.pdf [Accessed: September 28, 2012]

[7] Kiljunen T, Jarvinen H, Savolainen S. Diagnostic reference levels for thorax X-ray examinations of paediatric patients. The British Journal of Radiology. 2007;**80**(954):452-459

[8] Linet MS, Kim KP, Rajaraman P. Children's exposure to diagnostic medical radiation and cancer risk: Epidemiologic and dosimetric considerations. Pediatric Radiology. 2009;**39**(Suppl 1):S4-S26. DOI: 10.1007/s00247-008-1026-3

[9] Piepsz A, Hahn K, Roca I, Ciofetta G, Toth G, Gordon I. A radiopharmaceuticals schedule for imaging in paediatrics. Paediatric task group European association nuclear medicine. European Journal of Nuclear Medicine. 1990;**17**(3-4):127-129

[10] Smith T, Gordon I. An update of radiopharmaceutical schedules in children. Nuclear Medicine Communications. 1998;**19**:1023-1036

[11] Jacobs F, Thierens H, Piepsz A, Bacher K, Van de Wiele C, Ham H. Optimised tracer-dependent dosage cards to obtain weight-independent effective doses. European Journal of Nuclear Medicine and Molecular Imaging. 2004;**32**:581-588

[12] Lassmann M et al. The new EANM paediatric dosage card. European Journal of Nuclear Medicine and Molecular Imaging. 2003;**34**(5):796-798

[13] Treves ST, Davis RT, Fahey FH. Administrated radiopharmaceutical doses in children: A survey of 13 pediatric hospitals in North America. Journal of Nuclear Medicine. 2008;**49**:1024-2027

[14] Gelfand MJ, Parisi MT, Treves ST. Pediatric radiopharmaceutical administered doses: 2010 North American consensus guidelines. Journal of Nuclear Medicine. 2010;**52**:318-322

[15] Treves ST, Parisi MT, Gelfand MJ. Pediatric radiopharmaceutical doses: New guidelines. Radiology. 2011;**261**:347-349

[16] Fahey FH, Treves ST, Adelstein SJ. Minimizing and communicating radiation risk in pediatric nuclear medicine. Journal of Nuclear Medicine. 2011;**52**(2):1240-1251

[17] Zanzonico P. Nuclear Medicine Physics. Medical Physics. 2011;**38**(8):142-147. DOI: 10.1118/1.3611048

[18] Koizumi M, Takagishi H. The relationship between child maltreatment and emotion recognition. PLoS One. 2014;**9**(1):86093. DOI: 10.1371/journal.pone.0086093.Collection 2014

[19] Lassmann M et al. Paediatric radiopharmaceutical administration: Harmonization of the 2007 EANM paediatric dosage card (version 1.5.2008) and the 2010 North American

consensus guidelines. European Journal of Nuclear Medicine and Molecular Imaging. 2014;**41**(5):1036-1041

[20] Gelfand MJ, Parisis MT, Treves T. Pediatric radiopharmaceutical administered doses: 2010 North American consensus guidelines. Journal of Nuclear Medicine. 2011;**52**:318-322. Available from: http://www.imagegently.org/Portals/6/Procedures/Pediatric_dose_consensus_guidelines_Final_2010.pdf [Accessed: December 23, 2015]

[21] Lassmann M, Biassoni L, Monsieurs M, Franzius C, Jacobs F. The new EANM paediatric dosage card. European Journal of Nuclear Medicine and Molecular Imaging. 2007;**34**:769-798

[22] Lassmann M et al. The new EANM paediatric dosage card: Additional notes with respect to F-18. European Journal of Nuclear Medicine and Molecular Imaging. 2008;**34**(5):796-798

[23] Charnock P, Moores BM, Wilde R. Establishing local and regional DRLs by means of electronic radiographical X-ray examination records. Radiation Protection Dosimetry. 2013;**157**:62-72

[24] Vassileva J, Rehani M, Kostova-Lefterova D, Al-Naemi HM, Al Suwaidi JS, Arandjic D, et al. A study to establish international diagnostic reference levels for paediatric computed tomography. Radiation Protection Dosimetry. 2015;**165**(1-4):70-80

[25] Goske MJ, Strauss KJ, Coombs LP, Mandel KE, Towbin AJ, Larson DB, et al. Diagnostic reference ranges for pediatric abdominal CT. Radiology. 2013;**268**(1):208-218. DOI: 10.1148/radiol.13120730

[26] Society of Nuclear Medicine and Molecular Imaging (SNMMI). Nuclear Medicine Radiation Dose Tool. 2018. Available from: http://www.snmmi.org/ClinicalPractice/doseTool.aspx?ItemNumber=11216&navItemNumber=11218

[27] ARSAC. Notes for guidance on the clinical administration of radiopharmaceuticals and use of sealed radioactive sources, Adminstation of radioactive substances advisory committee. Nuclear Medicine Communications. 2000;**21**:S1-S93

[28] National Council on Radiation Protection and Measurements (NCRP). Reference Levels and Achievable Doses in Medical and Dental Imaging: Recommendations for the United States. Report No. 1722012. pp. 1-91